T4-AJF-614

E. M. CUDAHY
LOYOLA
UNIVERSITY
MEMORIAL LIBRARY

Electromagnetism, Man and the Environment

JOSEPH H. BATTOCLETTI, Medical College of Wisconsin

Electromagnetic pollution is the permeation of the environment with undesirable static and alternating electric and magnetic fields. The undesirable fields are usually man-made. Electromagnetic pollution is different from other types of pollution, such as air, water, and noise pollution, in two ways. First, it is almost always *invisible*, and second, there are sometimes *therapeutic* effects. This book is mainly concerned with the *effects of electromagnetism on the physiology of man*. The most prevalent sources of man-made electromagnetic fields are the generation, the transmission and distribution systems of electrical power, and the countless number of electrical devices which use electricity. The world's use of electricity has increased exponentially as the standard of living of man has improved. Specific sources of man-made electromagnetic fields are listed and described. Some of the obvious thermal effects on man, particularly from ultra-high-frequency and microwaves, are cited, as well as lesser known non-thermal physiological effects. Related to these are the therapeutic uses of electric and magnetic fields. In addition to electrodiathermy, electrotherapy is widely used in Eastern European countries, and electric fields have been used to promote bone regeneration in the USA. The controversy over the legitimacy of non-thermal electrotherapy is of great importance to the final judgment regarding the significance of electromagnetic pollution. A convenient method of specifying what is meant by low-level electromagnetic fields is the examination of human exposure standards which have been established by regulatory agencies. The book lists the standards the USA, Russia, and other countries have formulated, and discusses the significance of the large differences in allowable exposure levels. Finally, the book provides an overall discussion of the effects of electromagnetic fields.

Dr Battocletti is Project Director of Nuclear Magnetic Resonance in the Department of Neurosurgery at the Medical College of Wisconsin and co-inventor of the nuclear magnetic resonance flowmeter.

WESTVIEW ENVIRONMENTAL STUDIES
Editors: J. Rose (UK) and E. W. Weidner (USA)

Already published

Climate and the Environment

JOHN F. GRIFFITHS

Pesticides: Boon or Bane?

M. B. GREEN

WESTVIEW ENVIRONMENTAL STUDIES

Editors: J. Rose (UK) and E. W. Weidner (US)

ELECTROMAGNETISM, MAN AND THE ENVIRONMENT

Joseph H. Battocletti

WESTVIEW PRESS • BOULDER • COLORADO

QP
82.2
.E43
B37
1976

Westview Environmental Studies: Volume 3

All rights reserved. No part of this publication may be reproduced or transmitted in any form or by any means, electronic or mechanical, including photocopy, recording or any information storage and retrieval system, without permission in writing from the publishers.

Copyright ©1976 in London, England by Elek Books Ltd.

Published 1976 in London, England by Elek Books Ltd.

Published 1976 in the United States of America by
Westview Press, Inc.
1898 Flatiron Court
Boulder, Colorado 80301
Frederick A. Praeger, Publisher and Editorial Director

Printed and bound in Great Britain

Library of Congress Cataloging in Publication Data

Battocletti, J H
Electromagnetism, Man and the Environment

(Westview environmental studies ; v. 3)
Bibliography: p.
Includes index.
1. Electromagnetic fields—Physiological effect. 2. Electric fields—Physiological effect. 3. Magnetic fields—Physiological effect. I. Title.
QP82.2.E43B37 612′.01442 76-7905

ISBN 0-89158-612-1

Contents

Preface

When I was asked whether I would be willing to undertake the writing of a monograph on electrical pollution, I did not (even approximately) realise where it would lead me. I had just finished co-editing a book on *Biologic and Clinical Effects of Low-Frequency Magnetic and Electric Fields* (C. C. Thomas, Springfield, Illinois, 1974), subsequent to the co-programme chairmanship of a Symposium on the same subject in February, 1973. I *did* have a certain amount of knowledge and background on electrical pollution—enough to make a couple of presentations on the biological effects of microwaves.

I soon found that the subject of the biological and medical effects of electric, magnetic and electromagnetic voltages, currents and fields was dynamic and developing rapidly, not only in the scientific area, but in regulatory circles as well. Countless references were found on the many facets of these effects. Each reference sought out and studied led to many more related references. Thus, the library search would have mushroomed out of control, if a limit had not been placed on the scope of the book. This led to emphasising primarily the effects of electric, magnetic and electromagnetic *fields* on *humans*. Even with this limitation, there are still a large number of references. Those which are applicable to the thesis of this book and which were accessible were selected. These can be used as starting points for the reader to ferret out many other worthwhile and original papers. Emphasis has been placed on the use of the latest material available. Probably, by the time this book goes to press, there will be a number of important new developments.

Thus, this book on electrical pollution discusses how humankind is affected by the electric, magnetic and electromagnetic fields which exist by virtue of the present technological age. The major sources of these fields are delineated in Chapter 2.

I have tried to present both a dispassionate and objective appraisal of electrical pollution, except for the last part of the final chapter, in which I state my own personal viewpoint and conclusions. This approach has probably led to a certain degree of terseness and succinctness in the presentation of the subject matter.

I would like to thank Josep G. Llaurado, M.D., Ph.D., and Anthony Sances, Jr., Ph.D., for their encouragement and help which enabled me to undertake and complete this book. I also wish to acknowledge the patience and moral support of my wife, Rosemary, who has accepted a home office area (in the living-room of our home) cluttered with piles of books, magazines, and papers, which were used as background material. I would also like to thank J. Rose of Blackburn College of Technology and Design for the opportunity to write this book.

Joseph H. Battocletti, M.Sc., Ph.D., P.E.
May 15, 1975

1

What is Electrical Pollution?

Electrical pollution is the permeation of the environment (air, land, water) with *undesirable* static and alternating electric and magnetic fields. The undesirable fields may be of natural origin, but most often they are man made. Man has always been surrounded by the earth's magnetic field, the naturally occurring electrostatic fields and those fields created by lightning and wind, and by electromagnetic radiation generated by the sun. There is little doubt that these natural fields have had some influence on man. However, it has not been until the last several decades that concern has been expressed over man made electromagnetic fields from power, communication, ranging and electrical process systems. Truly, man is now immersed in a wide variety of man made fields at various energy levels.

Electrical pollution is different from other types of pollution such as air, water and noise pollution, in two ways. Firstly, it is almost always *invisible* and secondly, there are sometimes *therapeutic* effects. One can see, smell, feel and taste air and water pollution, and hear and feel noise pollution, all in disagreeable ways; but, one seldom senses electrical pollution and therefore most people are not aware of the presence of this type of pollution. Two major exceptions to invisibility are the electrical discharge and the warmth caused by the high-level radio-frequency and microwave fields used in heat therapy.

We are interested not in the effects on equipment (e.g. radio and television interference) of extraneous and unwanted electromagnetic signals, but in the effects on man. It is this that we call electrical pollution, or more generally electromagnetic pollution, since both electric and magnetic fields are involved.

2

Sources of Electrical Pollution

Man made electrical pollution probably began with the invention of the high voltage, low current electrostatic generator (Pieter van Musschenbroek, 1746) and the low voltage, high current primary battery (Alessandro Volta, 1800). However, their use was localised to research laboratories. The real onslaught came near the turn of this century when electric power was made available to the general public for lighting, transportation and motors. Then, electric and magnetic fields at the power frequency pervaded homes, factories, schools, offices, the streets, and most importantly, the power generating and transmission systems. Electronics was born in the early 1900s, introducing the potential for electric and magnetic fields in an almost limitless frequency spectrum.

In this chapter, we will attempt to list and describe specific sources of man made electric and magnetic fields which already exist in the world of today and also those which will exist in the world of tomorrow.

It has already been established beyond a shadow of a doubt that some sources produce extremely deleterious effects. It is also probably true that some sources produce no effects; in fact, some therapeutic effects are claimed. But there are many sources whose effects are as yet unknown. Some of the obvious effects will be mentioned in this chapter, but a more thorough examination of possible biological effects of electric and magnetic fields is made in subsequent chapters.

2.1 D.C. Magnetic Fields

2.1.1 Man-made Fields

In my work in nuclear magnetic resonance (NMR) measurements and applications, I have worked with d.c. magnetic fields of up to 12 000 gauss, and have needed, on occasion, to place my hands in the fields. In fact, in applications of NMR to blood flow measurements, the whole arm must be placed in fields of up to 1500 G for extended periods of time. I have used a 2000 G, 25 cm air-gap electromagnet, which accommodates the human head for cerebral blood flow measurements. I have observed no effects, and hope that there have been none.

Magnetic fields of up to 9300 G have been applied to the heads of patients to induce clotting in cerebral aneurysms by the injection of a suspension of iron microspheres suspended in albumin.[1]

Mice have been exposed to fields of up to 120 000 G in a water-cooled solenoid having an inner diameter of 2·5 cm.[2] Exposure times were from ten minutes to two hours. No ill effects from exposure to the magnetic field were observed for eight months after exposure for five of seven mice tested; one mouse died due to thermostress, and another died after four months from unknown causes.

Superconducting electromagnets are currently available in which the magnetic field is at an ambient temperature. Fields as high as 150 000 G are easily obtainable in almost any size of diameter field space in solenoid types. Squirrel monkeys have been exposed to such fields up to 100 000 G. It was found that the electrocardiogram (ECG) following the QRS wave was affected in a major way. However, cardiac activity was not affected.[3]

I have heard that an NMR scientist placed his head inside the active region of a cyclotron. He said he felt no sensations and reported no after-effects. However, the Stanford (California) Linear Accelerator Center has established safety standards for limits of magnetic fields to which humans are allowed to be exposed. These are: whole body or head—200 G for extended periods (hours), 2000 G for short periods (minutes); arms and hands—2000 G for extended periods, 20 000 G for short periods. These limits were established from the results of animal experimentation and careful observation of personnel at the Stanford Center. In Russia, however, A. M. Vyalov has recommended more stringent limits, and he also specifies field gradients. These are: whole body—300 G and 5–20 G cm^{-1}; hands—700 G, 10–20 G cm^{-1}.

So far, we have been talking about magnetic fields which are found in physical and biological laboratories. Relatively few people are exposed to these fields at the present time, although more of the general public may be exposed when more medical instruments using high-level magnetic fields are developed.

There are other sources of magnetic fields which exist in public areas, or which may exist in the future, that can affect greater numbers of the general public. They are generally of lower intensity than laboratory magnetic fields. Some of these sources are discussed below.

1. Magnetic detection systems are used: at airports, to frisk passengers invisibly; in libraries, to guard against illegal borrowing of books and magazines;[4] and in department stores, to minimise the losses due to shop-lifting. The last two systems produce much larger magnetic fields than airport detection systems. One airport detection system is the Westinghouse WD-2, which uses a magnetic field of 1·3 G (peak) at about 100 Hz,[5] (Marquette

University, where I have conducted a part of my library search for this book, has installed a magnetic detection system.)

2. Currents from batteries to power motors for electric automobiles, golf carts, lawnmowers, and portable tools, create d.c. magnetic fields.
3. High-voltage overhead and underground d.c. transmission lines produce large d.c. electric fields as well as magnetic fields. These are discussed later in this chapter.
4. Levitation by magnetic fields is being tested for future use in high-speed transportation systems. This source is discussed later in this chapter.
5. Magnetic separators are used in a variety of ways: overhead cranes to lift scrap iron; in recycling plants to separate ferrous metals from non-ferrous metals and glass; in the concentration of iron in low-grade ore;[6] in the removal of up to 60% of sulphur from pulverised coal;[7] in the purification of water and sewage by the removal of impurities attached to an iron oxide flocculent in a high-gradient magnetic field.[8]
6. Magnetic fields have been used to soften water, particularly for steam boilers.

2.1.2 Terrestrial and Extra-Terrestrial Geomagnetic and Geoelectric Fields

The subject of electrical pollution takes a peculiar twist when the subject of the geomagnetic field, terrestrial and extra-terrestrial, is considered. On the one hand, the effects of the earth's geomagnetic field fluctuations have been of some concern for a great many years, while, on the other hand, the effects of null magnetic fields in space travel have drawn the attention of the world's space scientists.

The strength of the earth's magnetic field is approximately 0·25 G near the equator, 0·6 G near the North Pole, and 0·7 G near the South Pole. It is believed that the earth's magnetic field is caused by the non-uniform rotation and cyclonic convection of the earth's molten nickel and iron core.[9] It is further believed that the field has reversed at irregular intervals of about 10^5 to 10^7 years in the past, and that the present direction has existed for about the last 10^6 years. Another theory hypothesises that the earth's magnetic field is created by electrical currents which continually flow between the earth and the plasma in which the earth is immersed. This requires that the earth possess a very large potential.[10] In fact, the charge theory has also been applied to the source of energy of the sun. The earth's magnetic field is decreasing at a rate of 0·00027 G per year, and the locations of the magnetic poles are slowly shifting in a northwesterly direction. Short-term rapid fluctuations also occur and are due primarily to magnetic storms and disturbances caused by solar activity; the magnitude of these fluctuations reaches 0·005 G.[11]

The physiological significance of the fluctuations of the earth's magnetic field has been a matter of conjecture and controversy for nearly fifty years.[12, 13] Statistical studies of death, nervous and psychic disorders, blood pressure, leucocyte count, rate of psychiatric hospital admissions, frequency of lymphocytosis, cerebrospinal meningitis, relapsing fever, etc., have been made to demonstrate the effect of geomagnetic fluctuations. Controlled experiments have also been made on a variety of life forms to prove their dependence on the geomagnetic field. Such experiments usually involved direction of motion or orientation of the life forms, including the homing instinct of pigeons and the migratory habits of other birds. It may therefore be expected that man made magnetic fields may also have some effect on these motions and orientations.

With the onset of manned extra-terrestrial flights, first around the earth and then to the moon, the potential effects of escaping from the earth's magnetic field (or any other magnetic fields) have been of concern. Experiments with animals and man in near-zero magnetic fields in the laboratory have demonstrated biological effects in simple life forms and small animals, but no effect on primates and man—with one notable exception:[13] there was a significant decrease (about 45% in the critical-flicker fusion threshold (CFF, the frequency at which a flickering light cannot be visually distinguished from a constant light'.[14, 15]

It is known that the earth possesses an electric charge potential. The normal average fair-weather (lack of thunderstorms) potential at the earth's surface is 130 V m^{-1} (negative).[11, 16] It decreases nearly exponentially with altitude, falling to about 1 V m^{-1} at 20 km. The electrosphere (upper atmosphere) carries a positive charge and potential. When there is thunderstorm activity, or when a dry wind blows, rapid and large field fluctuations may occur. Electric fields up to 3000 V m^{-1} have been measured. The earth's potential may even reverse, attaining levels as high as 1500 V m^{-1}. Geographically, the normal fair-weather potential is greater in central latitudes and decreases towards both the equator and the poles.

Solar magnetic disturbances are able to induce electric fields on the earth's surface; magnitudes up to 6 V m^{-1} have been measured. Where conductors are present, such as electric powerlines, substantial currents are generated by solar magnetic disturbances.[17] It is of interest to note that we are entering a period of increasing solar activity, which is expected to reach a maximum intensity around the year 1980.

Man is shielded from these electric fields by metal-frame buildings, automobiles, and aircraft. These serve as Faraday cages,* and correspond to the near-zero magnetic fields man encounters in space travel.

Although not geophysical, there are other important sources of environmental electrostatic fields. When two objects are suddenly separated, or when a piece of material is broken, it is found that one object is charged with electricity of one sign and the other with that of the opposite sign. Some examples are: rubbing a glass rod with a silk cloth; rubbing a stick of hard rubber with a piece of fur; rapidly moving belts over pulleys; rubber tyres of vehicles rolling over the surface of the ground; walking on carpet; the passage of dry wind over objects; the blowing of dust particles or water droplets through pipes. This type of static electricity is called 'frictional' electricity or 'triboelectricity'. Although not related to our specific interests, static electricity can be dangerous in the presence of flammable gases and liquids, such as petroleum products, cleaning solvents, liquid hydrogen, anaesthetics, and hydrogen balloons. Special precautions must be taken when working with these materials.

Another source of static electricity is due to the many types of plastics which are used in wearing apparel, furniture, curtains and drapes, and carpeting. Some plastic-type clothes acquire an excess of electrons, and are thus attracted to the body, which is normally at a positive potential. Other synthetic clothes, such as polyesters, nylon, and plasticised vinyl, give off electrons and become positively charged. One wonders what effect these artificially created electric fields have had and will continue to have on man. Some work has been done to attempt to correlate various physiological effects with electric field fluctuations, with some positive results.[18]

A certain amount of ionisation of air gases occurs. Normally, these ions are positively charged. Under some circumstances, it has been observed that the presence of negative ions in air brought about beneficial physiological and psychological effects to man.[19–21] The negative ions are produced by the proper application of artificial electric fields.

2.2 Power-Type Frequencies

The most prevalent sources of man made electric and magnetic fields are the generators, the transmission and distribution systems of electrical power, and the electrical devices which use electricity. As all countries of the world improve the standard of living of their citizens,

* Faraday cage—earthed wire screen which shields a piece of equipment from any external electric field.

the numbers and strengths of these sources will increase. The increase in the production of electrical energy in the USA and the USSR is illustrated in Figure 2.1. These are the largest users of electricity. Their combined production comprises nearly one-half of the world's total production. The use of a logarithmic scale for energy production illustrates that the rate of increase is decreasing for the USA, but is remaining the same for the USSR. The energy capacities of the USA, USSR and the world are shown in Figure 2.2. The rates of increase are nearly constant (except for the USSR, where the rate decreases after 1965). The combined capacity of the USA and USSR has been

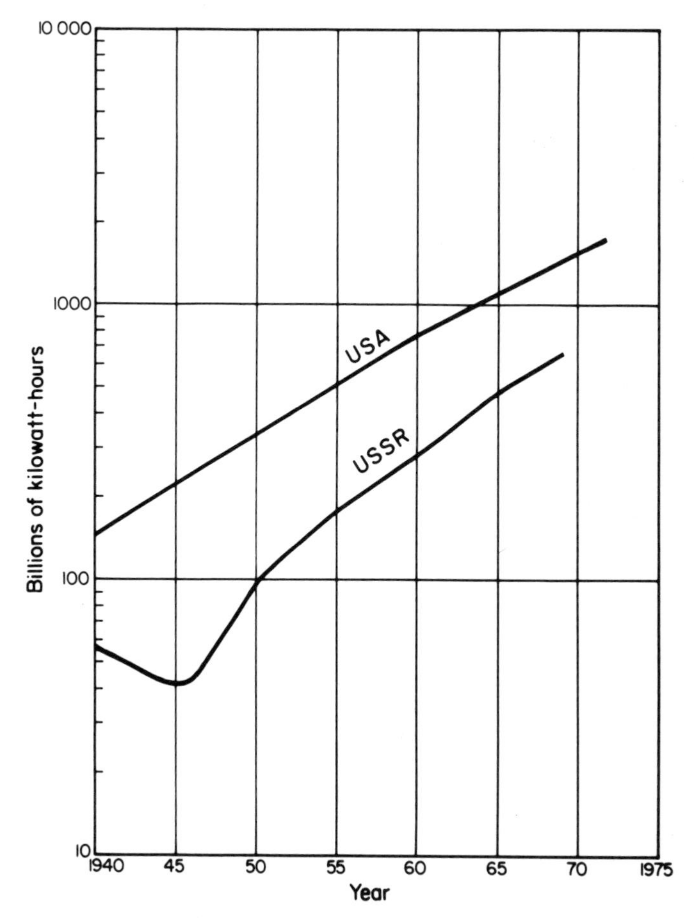

FIGURE 2.1 Electrical energy production from 1940–72 for the USA and USSR. A straight line is indicative of a constant rate of increase, since the vertical axis is plotted on a logarithmic scale. Data was obtained from various World Almanacs

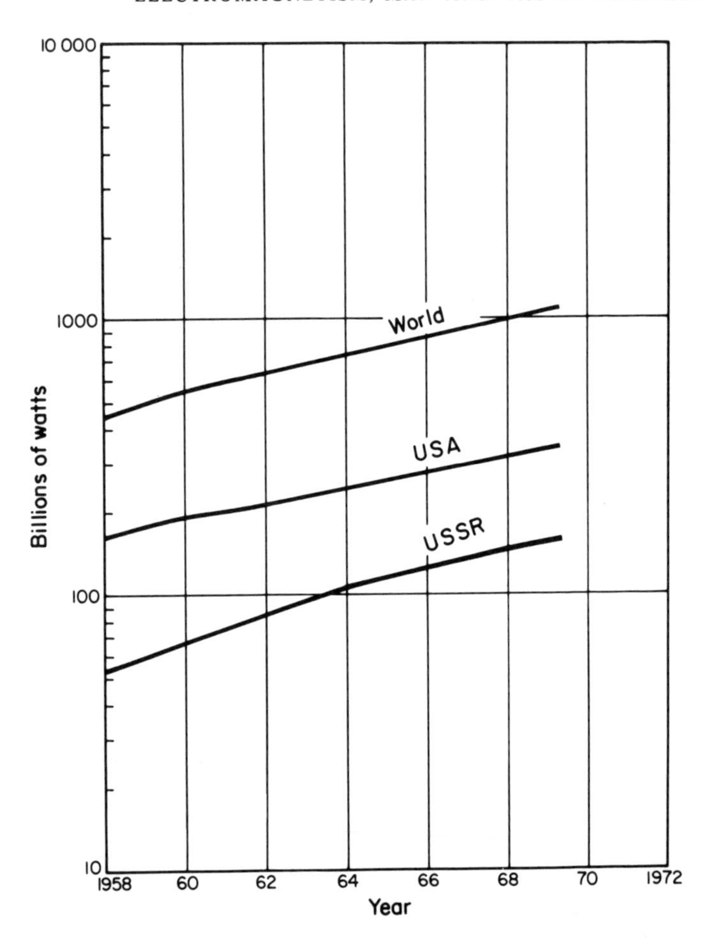

FIGURE 2.2 Electrical generating capacity (logarithmic graph) from 1958–69 for the World, USA, and USSR. Data was obtained from various World Almanacs

maintained between 46 and 47% since 1959. An increasing amount of electricity is being used in the home where people usually spend the majority of their time. In the USA, in 1966, 29·6% of the electrical power was used by residential consumers. This percentage had increased from 26·1% in 1955.

With the current emphasis on the limitations of oil and natural gas, added emphasis will be placed on electrical energy, generated from coal or nuclear energy. However, 'ecological' organisations have been formed to attempt to reduce the usage of all types of energy, even if this means a reduction in the standard of living. Also, definite programmes are being instituted by the power generating companies to reduce the wasting of electrical energy. These two latter movements may lessen the extent to which the use of electrical power increases in years to come.

2.1.1 Transmission Lines

An important source of electric and magnetic fields in the environment is the transmission line, both overhead and underground. At present, long-distance transmission of electric power occurs at extra-high voltage (EHV), operating at hundreds of kilovolts and 1000–2000 amperes. For these lines, the electric fields surrounding the transmission lines are of primary importance. Metallic objects may cause these fields to be increased to much higher levels in the vicinity of the metallic objects.

There is an increasing use of underground EHV cables, particularly over short distances, for various reasons, some aesthetic and others practical (such as for protection from the weather). Low voltage underground distribution systems are being introduced more and more at the final users of electric power, displacing the unsightly line of poles and wires. In the future (estimated to be beyond 1983), cryogenic (filled with liquid nitrogen or liquid hydrogen) and superconductive (filled with liquid helium) transmission cables will probably be used. For these lines, the currents will be in excess of 20 000 A. For example, a test section of superconducting cable has been built and tested; it operated at 345 kV and 1600 A.[22] For these systems, magnetic fields in the environment will be of importance.

There is a definite trend for transmission at higher and higher voltages. This results in higher electric and magnetic fields in wider corridors along the right-of-way. However, by using higher voltages, less overall right-of-way is required. For example, operation at 800 kV can deliver five times more power than at 362 kV and thirty times as much as at 130 kV.

For long-distance transmission, greater efficiency is obtained by operating at d.c. (direct current). A number of systems have been built since 1954. Two of the most important systems are:

1. Six hundred kilometres long, 500 kV, 600 megawatt (MW) lines to link the two islands of New Zealand. Forty kilometres of this is cable.[23]
2. Two 1350 km long, 800 kV ($\pm$400 kV), 1300 MW lines, called the Pacific Northwest-Southwest Intertie, in the USA.[24]

Other important systems have been planned in Russia and in Mozambique.[25, 26]

Modern a.c. power systems usually consist of the following: (1)

generating stations, (2) step-up transformer stations, (3) main transmission lines, (4) switching stations, (5) step-down transformer stations, (6) primary distribution lines, (7) service transformer banks, and (8) secondary lines to the final user.

Transmission at a.c. is generally made in a balanced three-phase, three-wire system. Each phase is shifted 120 electrical degrees from the other two phases. A sum of all the voltages gives a net voltage of zero. A concerted effort is made by the power companies to balance the currents in each line. This fact will cause the fields to be low at distances from the transmission which are appreciably greater than the wire-to-wire spacing.

There may be several primary distribution lines, each operating at successively lower voltages. This is particularly true for large systems in which the main transmission line is at EHV levels. Secondary lines are usually under 10 000 V. Electrical services to private residences are generally 120/240 V, three-wire, single-phase. Thus, one expects the extraneous electric and magnetic fields to decrease as one goes down the ladder of the electrical power system. However, as one does, the number of people affected increases at each rung of the ladder.

Where main-line transmission occurs at EHV d.c., after step-up transformation, a.c. must be converted (rectified) to d.c., and before step-down transformation at the receiving end, the d.c. must be inverted (changed back to a.c.). Rectification and inversion are accomplished using mercury-arc or thyristor devices. These devices cause another type of electrical pollution, radio frequency radiation, which we will discuss later in this chapter. This r.f. radiation occurs during switching periods of the electronic devices. To minimise this radiation, huge Faraday cages have been built for the Inverter Station in Sylmar, California at the south end of one of the 800 kV Pacific Intertie d.c. transmission lines. Each cage is made of 5 cm square steel mesh; it is 20 m high and has a 90×107 m base, and cost US $1 250 000 (when built). The Faraday cages (or Faraday shields as they are sometimes called) were used primarily to avoid interference with radio and television, but they also serve to reduce the impact of radiation on the environment.

Transmission at EHV d.c. is generally made on two wires, one positive and the other negative compared to ground. If the currents in the two lines are equal, the net field (electric and magnetic), at large distances compared with the line-to-line spacing, is low.

2.2.2 *Quantification of electric and magnetic fields*

Of concern to us are the magnitudes of the electric and magnetic fields surrounding the transmission lines, particularly the area below the lines.

Magnetic field The currents in the wires generate a magnetic field which is proportional to the current, and inversely proportional to the distance from the wires. Since the earth is mostly non-magnetic, both air and earth can be considered homogeneous from a magnetic viewpoint. (This is not true for the electric field, since the earth possesses non-zero conductivity, and the relative dielectric constant of earth is different from that of air. The method of images is used to take the earth's conductivity into account.) The magnetic flux density field lines form concentric circles round a long, straight, current-carrying conductor. The magnitude of the flux density vector is given by the equation

$$\mathbf{B} = \mu_0 I / 2\pi r \text{ tesla (T)} \quad (1)$$

where $\mu_0 = 4\pi \times 10^{-7}$ Henry per metre (H m^{-1})
r = radial distance from the current to the field point (metres)
I = current (Amperes)

[To convert **B** to the units of Gauss, the right side of Equation (1) is multiplied by 10 000.]

When there are multiple conductors, such as in the transmission lines we are considering here, **B** must be found for each conductor and vector summation made. For two-wire d.c. lines, the direction, and thus the sign, of I is different for each of the two wires. For three-phase, a.c. lines, the current (I) is alternating and can be treated in two ways: first, as time varying quantities shifting in time by one-third of a cycle from each other; or second, as phasor quantities, with phase angles of zero, +120°, and −120°. (These considerations of phase difference must also be taken into account in the calculation of electric fields.) The magnetic field at ground level beneath a typical overhead EHV line is less than one gauss (e.g., for a single conductor 25 m above ground, carrying 2000 A, tne field at ground level directly beneath the conductor is 0·16 G.) However, inside tunnel passages carrying high-voltage underground cables, the magnetic fields may reach levels of 20 G. For cryogenic EHV underground cables, the magnetic fields in cable passages will probably exceed 200 G.

Electric field The main electric field generated by overhead transmission lines is due to the potential diffierences between wires and between each wire and ground, metal towers, and nearby metallic objects, such as railroad tracks. This electric field is characterised by a high impedance, since its effect is due to capacitance coupling to objects immersed in it. Therefore, currents induced in conductive objects (animate and inanimate) are not very large, being below 100 μA. The perception threshold current ranges from 130 to 1000 μA for hand-to-hand or hand-to-foot pathways through healthy skin at power frequencies of 50 and 60 Hz. Both European and USA standards for maximum leakage current allowed for electrical appliances have been established at 500 μA.[27]

In our discussion, we are interested in the electric fields near ground and near the wires. An approximate equation for the electric field at the ground for a single overhead conductor is

$$\mathbf{E} = (V_0/k)[2h/(h^2 + x^2)] \text{ V m}^{-1} \tag{2}$$

where V_0 = potential of the conductor to ground (volts)
h = height of the conductor above ground (metres)
x = horizontal distance from a point directly under the conductor (metres)
$k = \log_e(2h/r_0)$, where r_0 is the radius of the conductor (metres)

For example, for $V_0 = 260$ kV, $h = 10{\cdot}35$ m, $r_0 = 1{\cdot}22$ cm ($0{\cdot}0122$ m), the electric field intensity at $x = 0$ is approximately $3{\cdot}38$ kV m^{-1}.

For multiple conductors, the field is determined by summing the contribution of all conductors, taking into account the phasic nature of the voltages, V_0, on each conductor. For a practical three-phase transmission line with overhead ground wires (Figure 2.3), the measured electric fields are slightly smaller than the value calculated

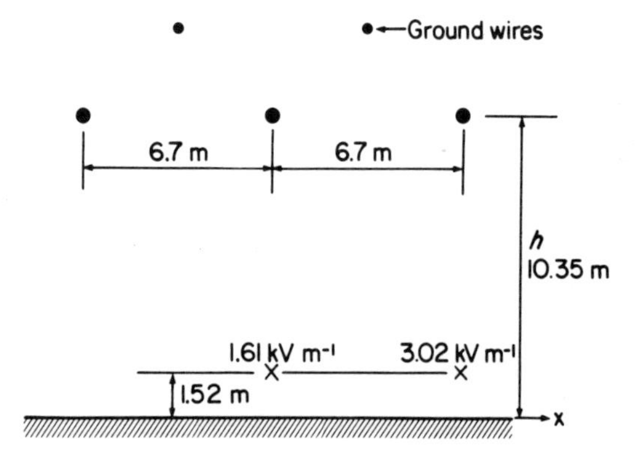

FIGURE 2.3 A typical three-phase EHV transmission line (260 kV from each line to the ground), showing the electric field intensity at two locations close to the ground

above for locations directly under the centre conductor, but very nearly the same under the outside conductor.[28] These levels should be compared with the natural levels of electric field existing on the earth's surface and in the earth's atmosphere. The accepted average value of the natural electric field is 130 V m^{-1}; it is greater in the central latitudes and decreases towards the equator and the poles.[16]

Once a transmission line is built, its maintenance is usually performed with the transmission line energised. Two techniques are preferred. In one method, a 'hot stick', about 3 m long, is used; in the second method, the linemen work 'bare handed' next to the energised conductor. An approximate equation for the electric field directly beneath a conductor at voltage, V_0, with respect to ground is,

$$\mathbf{E} = (v_0/k)\,(2h/(2hy - y^2))\ \mathrm{V\,m^{-1}} \tag{3}$$

where y = distance from the centre of the conductor (metres). The other quantities are defined in Equation (2). For the example treated above, the electric field at a distance 3 m below the 260 kV conductor is computed to be 13·6 kV m^{-1}; at the surface of the conductor, the electric field is maximum and is 2880 kV m^{-1}. A lineman will seldom be exposed to this maximum field level. For example, the electric field to which linemen using the 'hot stick' method of maintenance are exposed is about 70 kV m^{-1}, whereas it is 470 kV m^{-1} for those working 'bare handed' on a 345 kV system.[29]

Body currents have been measured in various positions and shieldings of linemen.[30] On the tower, the body current was 125 μA at a distance of 2·4 m from a conductor at 138 kV, and 395 μA at 3·2 m from a conductor at 345 kV. Significantly higher currents were measured when overhead shielding was not provided in the buckets used to lift the men up to the lines. It is estimated that a field intensity of 2·35 kV cm^{-1} (235 kV m^{-1}) will produce a current density of about 0·078 μA cm^{-2} entering the body.

Over a nine year period, 1963–72, ten linemen, some of whom worked on 765 kV lines, were carefully examined and followed by a team of physicians at The Johns Hopkins University, Baltimore, Maryland, USA.[29] An extensive variety of medical and laboratory examinations were made. Each lineman was examined seven times during the nine year period, including a session, each time, with a psychiatrist. There were no significant changes of any kind. The men remained essentially healthy in all respects. The psychiatrist could not detect any significant change in the emotional status in any of the men that could be related to the effect of work on the HV and EHV transmission lines. These results are to be contrasted with the results of a Russian study,[31] which mentions disorders of the functional state of the nervous and cardiovascular systems of people working in 400–500 kV switching structures.

The Russians have formulated standards and regulations for the protection of labourers during work on substations and transmission lines at EHV.[32] The lengths of time personnel are allowed to stay in various strengths of electric field during a 24 hour period are:

kV m^{-1}	*Minutes*
5	Unlimited
10	180
15	90
20	10
25	5

The generally accepted level in the USA is 15 kV m^{-1}. Barnes and Thoren[33] in a discussion of the future use of UHV (ultra-high voltage) transmission systems makes the statement that 'electric fields shouldn't be harmful'. They do not accept the work of the Russians, and believe that the use of a 5 kV m^{-1} limit is unrealistic, and that the 15 kV m^{-1} level is conservative.

In an a.c. system, current flowing in an overhead conductor causes a small electric field to be induced in the earth below the conductor. Even though this field is not large, the current passing through an object (such as an animal, man included), which contacts the earth in at least two places, can be significant (as much as, or greater, than current induced by the high-impedance field described above). This field exhibits a low impedance.[34] For example, suppose an animal makes good contact with the earth at two points; if its body resistance and the resistance of the earth between the two contacts are 1000 ohms, the current through the animal is approximately 30 μA if the electric field intensity in the earth is only 0·06 V m^{-1}.[35]

A third source of significant electric fields related to transmission lines is due to currents which flow into ground terminals. On many HV lines one or more neutral conductors (*see* Figure 2.3) is connected to ground periodically. Grounds are used at residential dwellings near the point where the electric power line enters the dwelling; the electric fields here are approximately 0·1 V m^{-1}, but may reach values 5–10 times this value.[36]

2.2.2 *Heavy Electrical Equipment*

The pieces of electrical equipment which probably have the largest electric and magnetic fields associated with them are the step-up and step-down transformer stations and the switching stations associated with HV and EHV transmission line systems. In these stations, unshielded bus bars operating at, or close to, transmission line voltages are required. Being closer to the ground than overhead transmission lines, there is a likelihood that a greater number of workmen and other visitors will be affected by the electric and magnetic fields present. The importance of this source is highlighted by Russian studies of the health status of people working in the electric field of open 400–500 kV switching structures.[31, 32] Electric field intensities of up to 27 kV m^{-1} have been measured in 500 kV switching substations.[37]

Electric and magnetic fields near industrial and institutional equipment (motors, transformers, generators, capacitor banks, switchboards) do exist. However, power lines to the equipment are often shielded by the metal conduits which carry them. Furthermore, the metal housing of the equipment provides shielding from the internal fields.

2.2.3 Household Appliances and Gadgets

The majority of people are exposed every day to the weak fields, both magnetic and electric, which are generated by household appliances and other electrical gadgets. Fields in the vicinity of many of these were measured as part of environmental impact studies for Project Sanguine, a high-powered extra-low-frequency (ELF) world-wide communication system under development in the USA. Some of the more significant data is presented in Table 2.1. Two items stand out head-and-shoulders above all others: first, the large electric field generated by an electric blanket, and second, the large magnetic field generated by a hair dryer.

TABLE 2.1

Magnetic and electric field values for some household appliances and electrical gadgets[36]

Electrical apparatus	*Magnetic field close to apparatus (gauss)*	*Electrical field 30 cm away (volts/metre)*
Hair dryer	10–25	40
Fluorescent desk lamp	5–10	—
Kitchen range	5–10	4
Electric shaver	5–10	—
Can opener	5–10	—
Colour television set	1–5	30
Food blender	1–5	—
Electric drill	1–5	—
Portable heater	1–5	—
Electric blanket	—	250
Broiler	—	130
Stereo radio	—	90
Refrigerator	0·001–0·01	60
Iron	0·01–0·1	60
Hand mixer	1–5	50

Both devices expose the user to these relatively high fields for long periods of time.

A new type of kitchen range has been marketed by the Westinghouse Electric Corporation, which cooks by magnetic induction. An oscillating magnetic field is generated by coils mounted just below the cooking surface. When an iron or steel pan is placed in the oscillating field, magnetic induction heats the pan, not the range surface. The Westinghouse CT-2 Cool/Heat Range operates from a 240 V, 60 A line.

Several generations have been exposed to the fields of household appliances for the greater portions of their lives with no apparent ill effects.

2.2.4 Electric Transportation Systems

(a) Electric-powered public transportation systems Electric-powered public transportation systems[38, 39, 40] may be placed in one of the following three classes:

Street, or trolley, cars operated at street level along main thoroughfares of cities and towns are fondly remembered by the writer as a single street car, with a single overhead trolley wire and two railroad-type tracks. I remember riding these in the 1930s and 1940s in a well-organised and coordinated transportation system on both sides of the Ohio River in Wheeling, West Virginia and Bridgeport, Ohio. They were still popular in the 1950s when I rode them daily to and from work in Chicago, Illinois. The 1960s saw the demise of street cars in most cities in the USA. There was a period, however, when trolley buses, running on rubber tyres with double overhead trolley wires, were used, but these, too, gave way to gasoline-powered and diesel-powered buses. However, these modes of transportation are still popular in other countries, such as Japan.

Elevated and subway trains and some trains operating on special surface right-of-ways in metropolitan areas, continue to be popular forms of public transportation in large cities, such as New York and Chicago in the USA, or Toronto and Montreal in Canada. For example, in New York, 8 000 000 people daily use public transportation comprised of subways, commuter railroads and buses. In some places, particularly in the USA, where the automobile became widely used, the smaller systems were discontinued. Today, however, with the necessity of reducing the number of automobiles travelling into the downtown areas of cities, new subway and elevated systems are being introduced. The most famous is BART (Bay Area Rapid Transit System) in San Francisco, California. Systems are also being built in Washington, D.C. and in Atlanta, Georgia, in the USA.

Surface railway systems operating over longer distances for both passenger and freight service are usually operated between two or more cities. There may be some subway and elevated sections, but these are only a small part of the whole system. European countries (Britain, Italy, Hungary, France, Belgium, Norway, Russia, Germany, Sweden, Finland, Czechoslovakia, Greece, and Austria) have many more electrified systems of this type than the USA. Expansion slowed substantially with the introduction of the diesel-electric locomotive. However, with the recent realisation of the limitations of fuel oil, there is renewed interest in fully-electrified transit systems.

Being concerned with the electric and magnetic fields which are generated by electric transit systems, we must consider several things:

How is the electricity supplied to the transit system?

What transformation takes place at transit system substations?

What is the level and type of electricity supplied to the electric locomotives?

How is the electricity supplied—in overhead wires, or by a third rail?

Answers to these questions lead to knowledge of the electric and magnetic field exposure levels for railway workers, the travelling public (in motion and at stations) and the residents who live near the railways.

Traction motors are powered primarily by direct current or single-phase low-frequency alternating current. In the USA, 25 Hz is used, whereas in Europe $16\frac{2}{3}$ Hz (one-third of the industrial frequency is obtained by frequency changers) is used. The low frequency allows single-phase alternating current traction motors to be used without the need for a complex system of speed reduction gears between the motors and the axles, since motor speed is less at the lower frequency. For d.c. traction systems, rotary converters used to be used to obtain direct current to power direct-current traction motors, but these were expensive. However, with the advent of mercury-arc rectifiers, thyristors, and silicon rectifiers, more d.c. traction motors are now used because of the better speed control attainable and the smoother acceleration. Also, the use of the common power frequency (60 Hz in the USA and 50 Hz in Europe) is now possible, since it is rectified either at substations or at the locomotive, to obtain direct current. Thus, frequency changers which are expensive and have higher operating and maintenance costs, are not needed. This fact, in itself, has given new life to electric transit systems, since electric power grids are available and are becoming more widespread as more and more generating stations and transmission systems are being built. This would result in lower electric energy rates and less investment and maintenance of transmission lines by the transit companies.

Therefore, power supplied to the locomotives may be either single-phase alternating current or direct current, which may be applied via overhead catenary wires or via a third rail. If it is supplied by overhead wires, the voltage is likely to be single-phase above 10 kV, although some overheads carry 3000 V d.c.; if by a third rail, the voltage is generally about 600–1200 V d.c. Table 2.2 gives some of the most common combinations of frequency and voltage and the method of application. In those cases where overhead catenaries supply the industrial power frequency (50 and 60 Hz), rectification by thyristors is used to obtain direct current for d.c. traction motors, except for Hungary, where a special type of synchronous motor-generator is used to utilise 50 Hz power directly.

TABLE 2.2

Electric transit systems. Frequency and voltage and method of application

	Frequency	*Overhead catenary*	*Third rail*
Alternating current	$16\frac{2}{3}$ Hz	15 kV Switzerland, West Germany, Sweden	
	25 Hz	11 kV USA	
	50 Hz	16 kV Hungary	
		25 kV France, Czechoslovakia, Romania, Russia	
	60 Hz	11 kV Capability in USA	
		25 kV Japan	
Direct current		3000 V USA, France, Belgium, Italy	600 V USA, West Germany
		1500 V Great Britain, Holland, Japan	660, 750 V Great Britain
			1000 V USA (BART)
			700 V USA (Washington, D.C.)
			1200 V West Germany

The magnetic field around a particular section of railway line exists only when current is flowing, which occurs only when a train is travelling on that section of line. However, the electric field exists at all times, since the potential is always present. The electric field of the overhead-feed system extends over longer distances than that of the third-rail system. In the third-rail system, the field exists primarily between the third rail and the ground, whereas in the overhead system it exists between the overhead conductor and the ground. Equations 1, 2 and 3 on pages 11, 12 and 13 can be applied here. Thus, from an electrical pollution standpoint, the overhead system contributes the most. The people who are awaiting the arrival of trains at stations would probably be most affected by these fields.

Electric power is usually supplied to the transit companies by local utility companies at a voltage level most convenient to the utility company. For example, the Pennsylvania Central Railroad (USA) is supplied at 13 200 V, 25 Hz, single phase by four utility companies. The railroad company increases this to 132 kV for distribution along its electric railway system. At substations, this is transformed to 11 kV for the overhead catenaries. In Hungary, a 110 kV transmission line supplies the electrified rail line from Budapest to the Austrian border. Substations transform this to 16 kV for the railway's overhead system. The Bay Area Rapid Transit (BART) system in San Francisco, California is supplied at 34·5 kV. At substations, this is transformed and rectified to provide 1000 V d.c. for BART's third-rail system. Sweden's extensive electric transit system is supplied at 6 kV, 50 Hz, three-phase. At substations, this is transformed and the frequency changed to 15 kV. $16\frac{2}{3}$ Hz for its overhead catenaries. Electric and magnetic fields exist around these distribution lines and in the railway substations, as discussed earlier in this chapter.

(b) Magnetic levitation and propulsion High-speed ground transportation at speeds in the range of 400–500 kilometres per hour requires an almost non-contact suspension system. Suspension, or levitation, by means of magnetic fields has become the favoured system within the past seven or eight years. In addition to suspension, magnetic fields are used for propulsion. Although designed for high speed, magnetic suspension and propulsion may eventually find application in lower-speed systems as well.

The Rohr Industries, California (USA) demonstrated and operated, at moderate speeds, a magnetically guided and propelled vehicle at the 1972 International Transportation Exposition. The system, called ROMAG®*, used linear induction electric motors on one side of the vehicle to develop both thrust and dynamic active suspension.[41, 42] Suspension was accomplished by *ferromagnetic attraction* between a laminated steel guideway rail and the vehicle magnet, which was excited by multi-phase windings. The vehicle magnet rode under the guideway rail. The guideway rail contained short-circuiting conductors of aluminium so that the rail operated as a rotor of a squirrel-cage induction motor, which gave the vehicle its forward thrust. Braking was accomplished by introduction of a direct current into the excitation windings.

In West Germany, two companies, Krauss Maffei and Messerschmitt-Bolkow-Blohm, working with the German Research Agency (Forschungs Gemeinschaft) have both built and demonstrated vehicles which also accomplish levitation by *ferromagnetic-attractive*

* ® Registered, or Trademark.

forces.[42] The vehicle lift magnets are attracted to the bottom of the inverter U-shaped rails along each side of the guideway. Propulsion is accomplished separately by means of a vertical aluminium reaction rail in the centre of the guideway, which operates as the secondary of a linear induction motor.

In Japan,[42, 43] superconducting magnets are used to obtain suspension and propulsion in a different way from the method described above. Suspension is accomplished by a *repulsive force* interaction between conducting coils in the roadbed and superconducting coils installed in the vehicle. Propulsion is obtained by another set of coils, one in the roadbed (the armature) and the other (superconducting field coils) in the vehicle, acting as a linear synchronous motor. A vehicle of this type has been successfully tested by the Japanese National Railways in conjunction with Hitachi, Mitsubishi, and Toshiba.

The Wolfson Foundation, in Britain, is funding three universities to investigate different methods of magnetic levitation. The University of Warwick intends to construct a test vehicle that will run above aluminium tracks and be levitated by superconducting magnets. The University of Sussex will study levitation using conventional electromagnets above iron rails in a system utilising magnetic attraction. The University of Wales Institute of Science and Technology will be looking at other magnetic levitation systems including repulsion by permanent magnets on both vehicle and track.

The European Common Market is studying a transportation system which will be powered by linear-propulsion motors and probably levitated by air. The proposed speed is 300 km h^{-1} and would link major cities in Europe.

It is estimated that in a levitated vehicle employing superconducting magnets, the magnetic field at the floor of the passenger cabin may reach levels of 300 G (0·03 T).[44] It is proposed to decrease this by active shielding to about 50 G by producing an opposing field. The level of fields to which maintenance personnel are exposed is probably larger than these values.

(c) The electric car It is anticipated that the world's supply of petroleum will be near depletion sometime before the year 2050. The USA supply will probably be depleted by the year 2000. Personal transportation by automobile will probably survive by the use of alternate sources of energy. One solution of this problem has already been undertaken, namely, the electric-powered car. (It is interesting to remember that, in the early days, the electric automobile was used before the sudden rise of the internal combustion engine.)

The battery will be the source of power for the electric-powered automobile, and will need to be recharged, probably nightly, to maintain full charge. As a result, the electric utilities will need to deliver more energy to supply this increased demand. There are about 9·3 kilowatt-hours (kW h) of energy contained in one litre of gasoline. It has been estimated[45, 46] that when the electric automobile is widely used, by the year 2000, the portion of the USA utility capacity used for charging will be about 15%. Fortunately, most of this charging will occur at night-time when the demand by home and industry is far less than during the day-time.

With regard to added electric pollution from the electric automobile, two aspects should be considered. First, there will be increased fields from transmission lines, switching substations, and distribution systems. Second, since the entire energy to operate the automobile is electrically derived, the fields, primarily magnetic, around the automobiles may be substantial.

2.3 Project Sanguine

Sanguine[47] was the name given to a proposed long-range project by the US Navy Department to develop a reliable and survivable communication system. The system would have transmitted high priority command and control messages, almost world-wide, to submerged submarines and other US forces, from a single underground transmitting location in the USA.

The Sanguine Project was probably the first technological development to be attacked by the general public because of the possible effects of electric and magnetic fields on the environment.

The antenna, between 480 and 2560 km of insulated cable, was to have been buried underground. The total power required to drive 150 A through the antenna was 20–30 MW. The antenna would have excited electromagnetic waves in the spherical cavity bounded by the ionosphere and the earth's surface. The dipole antenna was to be fed at the centre; current would return from the ends of the antenna after deeply penetrating the earth's crust. The electric field which would have been established would be principally vertically polarised, with a small horizontal component. It is the horizontal component which can be propagated downwards in sea water to communicate with submarines. The frequency of the signal would have been under 100 Hz.

The method has been tested several times on a small scale.[48] In 1963, a message was transmitted from a Test Facility in North Carolina to a submerged submarine using less than half a watt of radiated power. From a 23 km long atenna system at a test facility in Wisconsin (in 1971 and 1972), messages were picked up by a ground receiver in

Norway and by a submerged submarine, again using only half a watt of radiated power. Both antenna systems were above ground.

In spite of the apparent future success of the Sanguine Project, environmentalists and ecologists have been successful in delaying and virtually killing the project. Strong pressure was brought to bear upon the US Congress to withhold finances for the project. Owing to local opposition, it was difficult to find a suitable site for the transmitting facility. To obtain satisfactory antenna efficiency, the substrata of the earth must have a low conductivity. This limited the possible locations to the States of Texas, Wisconsin, New York, Michigan, North Carolina and Virginia. No State was anxious to have the facility built within its boundaries.

An environmental compatibility assurance programme was undertaken in 1968 to consider the project's impact on man's social environment and on organisms, as well as to study its ecological implications. The following conclusions were made:[49, 50]

1. No substantive evidence had been found in the literature to suggest significant reactions at Sanguine levels (0·07 V m^{-1} and 0·20 G).
2. There were no basic mechanisms described for biological reactions at Sanguine levels.
3. Electromagnetic fields similar to those associated with Sanguine now exist with no obvious effects.
4. Research to date indicated that intensities much higher than Sanguine levels were required for any type of evident reactions.
5. Ecological surveys at the Wisconsin Test Facility demonstrated no adverse effects. The normal ambient electric field values due to the power frequency (60 Hz) was 0·040 V m^{-1} (maximum) and 0·003 V m^{-1} (mean of 200 sampling points). These values should be compared with the electric field in the vicinity of EHV transmission lines, which can reach levels of several thousand volts per metre (*see* page 12). The electric and magnetic fields for home appliances exceed those expected for the proposed Sanguine project (*see* Table 2.1).

An extensive medical and laboratory evaluation of the health of personnel working near the Wisconsin Test Facility from 1971 and 1972 was made and compared with a control group.[50] The following areas were taken into account: constitutional symptoms, endocrine systems, neuromuscularity, ophthalmological, cardiovascular, haematalogical, psychological, and integration of physiological function. No significant differences were noted between Sanguine and control subjects, which parallel the findings by Singewald, *et al.*[29] on linemen working on high-voltage transmission lines.

In spite of all the data which have been compiled to demonstrate the safeness of Project Sanguine, there has been a fearful hesitation on the part of the public to accept this supposed pollution of the environment. On several occasions the project was thought to have been killed; but, like the proverbial cat, it seemed to have nine lives. While the Sanguine Project rests in limbo, another version of the Extremely Low Frequency (ELF) Communications System, called the SEAFARER, has been authorised by the US Department of Defense. It will be installed either at Nellis Air Force Base in Nevada or at the Fort Bliss (Texas)–White Sands (New Mexico) military complex, or possibly in northern Michigan, where present studies are now being made.

2.4 Radiated Electromagnetic Energy

In the realm of radiofrequency radiation, awareness of possible harmful effects has existed from the early days of high-power radio transmission when temperature increases were observed in objects immersed in the near field of formal and informal antennae. Since then, this energy has been harnessed for many useful industrial and medical applications, such as induction and dielectric heating and electrotherapy. Induction heating is used in melting, forging, annealing, surface hardening, brazing and soldering operations. The age of the transistor would probably not have been realised without the growing of nearly perfect crystals of germanium and silicon in induction heating furnaces. Dielectric heating is used to speed up the gluing of plywood and plywood products, to facilitate the moulding of plastics, to expedite the vulcanisation of rubber, and to join or seal thin plastic materials. High-frequency heating has also been used in the sterilisation and preservation of foods. Therapeutic treatment of internal muscle aches and pains of humans by radio diathermy is well known.

The effects of high-power radar are well known; there have been reported fatalities. In a US Government report published in 1970,[51] numbers and types of exposure to non-ionising radiation are listed. The radiation accidents listed therein occurred during the period 1950–70. Seventy-eight incidents affecting 100 people were reported. By far the greatest number of injuries occurred to the eyes, and were due to occupational rather than non-occupational exposures. Microwave radiation was the primary type to cause injury. Today, there is controversy over the safety of microwave ovens, which are being used increasingly in the home. However, it should be pointed out that controlled use of microwaves has been beneficial both in industry and in medicine.

Weinberg[52] speculates that patients connected to equipment having a direct electrical path to the heart may be endangered by local application of diathermy or by being placed in a strong radio transmitter field or radar field, not directly, but indirectly as follows: local rectification of the radio frequency may occur, and 60 Hz noise superimposed on the carrier, or the direct current resulting from the rectification, may cause ventricular fibrillation (defined as unco-ordinated asynchronous contraction of heart muscle fibres producing no pumping action on the blood.

In this section, we wish to discuss as sources of electrical pollution not only those which produce thermal effects, but those many other sources of radio frequency electric and magnetic fields in which many people are immersed.

2.4.1 Radio Frequency

(a) Radio and television broadcasting In 1971, there were about 8000 broadcasting stations in the USA alone, comprising about 900 television, 2700 FM, and 4400 AM stations.[53] The effective radiated power (ERP) of AM stations range from 100 W to 50 kW, except for the Voice of America short-wave stations which operate as high as 1 MW. The field strength of a 50 kW AM radio station is approximately 1 V m^{-1} at a distance of 1·6 km from the broadcasting antenna. The ERP for FM and TV broadcasting stations is generally greater than AM broadcasting stations because of the shorter distances over which the VHF and UHF waves propagate. The field strength at a distance of 1·6 km is about 1 V m^{-1} at ground level, although it may be significantly larger in line-of-sight with the radiating antenna, for example, in windows of tall buildings near the antenna. Close to the antennae (160 metres) field strengths may reach 40 V m^{-1}. Translating this into power density in free space, we find that 2 V m^{-1} is equivalent to about 1 μW cm^{-2} and 100 V m^{-1} is equivalent to about 30 μW cm^{-2} which is well below the established limits of exposure, except in eastern European countries.

A study of the intensities due to multiple radio, TV, and radio stations was made in the Washington, D.C. (USA) area in 1971.[54] The maximum power density observed was in the 1000–3000 MHz frequency band and was 7·7 μW cm^{-2} at the National Airport. On the basis of this study, it was recommended that no further broad-band environmental surveys be carried out unless prior calculations indicated their need. However, individual stations should be closely monitored to determine exposure levels at populated sites in the proximate area.

Measurement of electromagnetic radiation levels from selected transmitters operating at between 54 and 220 MHz was made in the area around Las Vegas, Nevada (USA).[55] The largest density of radiation measured was 0·813 μW cm^{-2}, an order of magnitude less than the maximum density measured in Washington, D.C. study.

Antennae were usually constructed in rural areas, away from centres of population. However, as the cities extended their boundaries, and as suburban living became more popular, homes and businesses began to move closer and closer to these antennae. Many homes and businesses are now located adjacent to the property lines of the antenna installations. Thus, more people are being exposed to higher levels of radiation than before. However, the level is probably less than 1 V m^{-1}, unless strips of metal in the buildings are functioning as parasitic elements of the radio station antenna, in which case, the levels may be in excess of 1 V m^{-1}. It may be particularly hazardous to locate hospitals and clinics near these installations if the effects Weinberg describes[52] occur.

In addition to the broadcasting transmitters, there were 300 000 amateur, 2 900 000 citizen band, 200 000 aviation service, 1 700 000 industrial service, 500 000 transportation service, 200 000 marine service, and 700 000 public safety service transmitters in the USA in 1970.[56] Although radio amateur operators in the USA are limited by law in the amount of power they can transmit (1000 W), they are very close to the source of radiated energy. Of course, most of the power is normally transferred along the transmission line to the antenna. However, there is radiation directly from the final tank coil and from the transmission line, if they are not shielded. Even if only 1 W is radiated, the field strength at a distance of 1 metre is about 5 V m^{-1}.

Recently, a novel method of increasing the range of radio and TV signals has been demonstrated by the Stanford Research Institute in Menlo Park, California (USA). A region of energised ionospheric electrons is formed in the ionosphere by radiation from a 500 kW short-wave radio transmitter on the ground. The energised region is 160 km across by 16 km thick, and allows reception of VHF transmission more than 1600 km from the source of transmission.[57]

(b) Induction heating The heating of conductive materials by means of eddy current losses induced in the material is called *induction heating*. It has been used since the early 1940s. Radio frequency current is passed through a coil, usually a solenoid. The material to be heated is placed through the lumen of the coil. Even though the purpose of the induction coil is to heat the material inside the coil, a substantial amount of energy is radiated into the surrounding space. Often the heating coil is prominently placed so that the heating phenomenon may be observed. This was true in several electronic induction heaters used to purify intrinsic semiconductor rods at a development and manufacturing company at which I worked in 1958.

Power to the coil may be supplied from a number of sources, the most modern of which is electronic. These sources are: 1. directly from a.c. mains (50, 60, 400 Hz); 2. from a motor–generator set (25–700 kW at 960, 3000 or 9600 Hz); 3. from a spark-gap converter (1–50 kW at 50–200 kHz); 4. from mercury-pool frequency converters (250 kW at 1000 Hz); and, 5. electronic (1–200 kW at 300 kHz up to the megahertz range). At the high powers, the current through the coil is hundreds of amperes.

In 1958, I made comprehensive tests of the radiation throughout a semiconductor research and development laboratory in which there were four R.F. induction generators. For two of them, the intensity near them exceeded 1 V m^{-1}. For the third, the intensity was in excess of 0·5 V m^{-1}, and for the fourth, the intensity was about 0·1 V m^{-1}. Furthermore, it was determined that not only is the R.F. radiated directly, falling off according to the inverse square law, but it is also propagated indirectly, along power lines, conduit and water pipes. For example, the intensity near a water pipe about 30 metres from one of the induction generators (12·5 kV A, 5·1 MHz) was 0·5 V m^{-1}.

(c) Dielectric heating Non-conductive materials may be heated by a process known as 'dielectric hysteresis', when the non-conductive materials are placed between the two conductive plates of a capacitor. The loss factor of the dielectric causes heat to be generated when the capacitor is excited by a radio-frequency voltage. Frequency may be in the range of 2–40 MHz (13 MHz and 27 MHz are commonly used), with power levels of the order of 10–200 kW. The voltage across the plates generally does not exceed 15 000 volts, and the gradient between the plates is between 100–2000 V cm^{-1}.

Because the size of the plates of the capacitor is usually much larger than the spacing between them, there is not much leakage of radio frequency fields. However, it is possible to obtain overdoses of direct radiation from the edges of the plates. The leakage from the coils, which make up the R.F. amplifiers can be minimised by shielding them by metal cabinets.

I am aware of a small company whose sole business is the sealing of plastic products, using dielectric heating at 27 MHz. Six or seven heating units are normally in operation at one time. Obviously, the employees are exposed to high levels of R.F. radiation. Moreover, no precautions had been taken to limit the range of the effects of direct and indirect radiation, since high levels of interference at 27 MHz have been observed by other businesses in the same building.

(d) Kirlian photography There is a pattern of coronal discharges or auras emanated from various parts of the human body. In 1939,

Semyon D. Kirlian and his wife, Valentia, used an electrotherapy machine to photograph the discharge patterns formed around the contact points of high-frequency voltage electrodes and various parts of the body touched by them. Today, a number of groups of scientists and non-scientists throughout the world are studying these discharges.[58] In fact, a company in the USA is marketing a 'Kirlian' Electrophotography Research Kit (The Edmund Scientific Company, Edscorp Building, Barrington, New Jersey 08007). The colourful haloes of blue, pink and crimson are likened to St Elmo's fire observed at the tips of pointed structures and to corona discharges around high-voltage transmission lines and apparatus. It appears that the intensity, colour and character of the patterns depend on the mental and physical state of the subject. Proponents of the Kirlian phenomenon conjecture that living organisms possess some form of an energy field which, in the case of humans, depends on physiological and psychological parameters. They believe that such dependence may be used as a tool to diagnose illness before the conventional symptoms appear. Other possible applications include the potential to explain acupuncture, psychic phenomena such as extrasensory perception (ESP), psychokinesis and healing by 'laying on of hands'. Finally, Kirlian advocates believe that by the judicious application of energy, similar to those normally emanating from humans, various diseases may be cured. More will be written about electro-therapy, a related subject, in a subsequent chapter.

In Kirlian photography a narrow pulse high-voltage radio frequency (ordinarily less than 1 MHz) is usually generated by a Tesla coil. The current passing through the biological substance is very small due to the small capacitance between the object and the source, and is considered by the Kirlian researchers to be harmless.

(e) Man as an antenna? Would you believe that it has been suggested that man could serve as a radio antenna? Experiments have been performed by Kurt Ikrath of the US Army Electronics Command, Fort Monmouth, New Jersey with human antennae at 4·3 MHz and 8·25 MHz at an input power of 1 W.[59] He reported that the body corresponds to a 1·2 metre long whip antenna at 4·2 MHz, with highly directional characteristics. While using man as an antenna may not be useful, it was suggested as long ago as 1904 that living trees could be used as short-range high-frequency antennae.

(f) Radio frequency diathermy Diathermy is the use of artificially induced local heat for therapeutic purposes. I can personally attest to its effectiveness, having had radio frequency diathermy applied to my back to relieve a severe muscle strain. Just prior to the application of this source of internal heat by induction (similar to induction heating described above), the pain was almost unbearable; afterwards, the pain was relieved.

R.F. diathermy machines usually operate at 27 MHz. When in the microwave region, 915 MHz and 2450 MHz are used in the USA.

Obviously the levels of radio frequency energy used for this beneficial therapeutic purpose is sufficiently high to produce a temperature rise inside the body. In two USA and two German diathermy units, the typical power density at a distance of 15 cm can be 100 mW cm^{-2}.[60] It is not known if this type of radiation is destructive to cells. It is known that certain parts of the body (the eye and the testes or ovaries) are particularly susceptible to radiation, and care must be taken to avoid exposing these organs to radiation.

(g) High frequency switching regulators High frequency switching regulating power supplies are being used more and more in electronic equipment because of greater efficiency and lighter weight. The line frequency supplies energy to a power transistor switch, which operates at 20–30 kHz. The high frequency voltage is transformed to the desired level, rectified, filtered and regulated. Direct current power of up to 10 kW has been generated in this manner.

There has been concern by the Federal Communication Commission (USA) about the EMI (electromagnetic interference) generated by these power supplies. Extremely effective shielding is required to eliminate EMI. Such shielding eliminates the possibility of adverse biological effects to man. However, maintenance personnel may be exposed to high levels of radio frequency radiation since they usually work on electronic equipment when the shields (cases, cans, etc.) are removed.

(h) Arc welders Arc welders radiate a wide variety of radio frequencies as well as light.

2.4.2 Microwaves

The effect of microwave radiation on biological systems has been the subject of numerous papers, articles, conferences, symposia, and *ad hoc* committees. The deleterious effect of radiation from microwave ovens has been discussed in some publications of consumer groups. Governments have passed laws to regulate the manufacture and operation of electronic equipment. For example, Public Health Service Act, PL 90–602, entitled the 'Radiation Control for Health and Safety Act of 1968', was passed by the US Congress to help protect the public from the dangers of electronic-product radiation, both the ionising and non-ionising types. Microwave radiation was one of the main constituents of non-ionising types. In this section, we will discuss various present and potential sources of microwave radiation which can be harmful to man.

(a) Radar One of the oldest applications of microwaves is radar. Because of the newness of microwaves and the ignorance of their possible effects on man, there were radiation accidents. It took many years before the problem came to be widely recognised. The primary source of danger to technicians occurs in the immediate vicinity of radar antennae. The bulk of the injuries have been to the eyes. It is possible to produce the same effect by looking back into the end of the wave-guide which feeds the antenna. Lesser radiation fields exist due to leakage from the various microwave components and in line-of-sight of radar antennae.

In the beginning, radar was used in the Second World War effort. Today, it is being used in a larger number of public and private places. Nearly every airport has radar to assist in air-traffic control. Ships employ radar to assist them in avoiding collisions with other ships. Shore-based radar is being used in San Francisco Bay (California, USA) to give protection from collisions and groundings; information is relayed to ships by radio. A radar chain covering the entire Westerschelde estuary in the Netherlands is being constructed. The frequency is 10 000 MHz, the peak power is 30 kW, and pulse repetition rate is 3000 Hz.

By 1969, there were nearly 3000 fixed radar devices in the USA.[61]

There have been a number of proposed electronic highway systems to allow high-speed and automated transportation; radar is one of the techniques to be employed.[62, 63, 64] Whereas the electronic highway is still far in the future, the use of radar to avoid automobile collisions by providing automatic braking is already being tested. Radar is also being tested in microwave speedometers and electronic-type licence plates which can provide a wide range of functions.[65] Another use of radar is by the police departments of many communities to measure the speed of highway vehicles to discover infringements of the speed limit. All these applications of radar, though at low power levels, will affect a greater number of people. A radiation of 50 mW yields a power density of 100 μW cm^{-2} at a distance of 5 metres.

The US Air Force is seeking a self-contained, small-area ground radar with a range of 100 metres to serve as a surveillance system around flight lines and other key facilities.

(b) Industrial applications Microwaves are used for a wide variety of industrial applications. Some of these applications are in the following areas: food, forest products, rubber and chemicals, plastics, and agriculture. Frequency is either 915 MHz or 2450 MHz; power levels range from 1–200 kW. The equipment usually has a conveyor belt to pass the material into and out of the working region. The entrance and exit ports are the primary sources of microwave leakage and thus of potential danger to employees. The second source of leakage is the hinged and sliding doors used to obtain access into the working region of the device for cleaning. In a typical investigation,[66] the maximum leakage power density 5 cm away from the conveyor belt entrance and exit slots varied from 1 mW cm^{-2} to 70 mW cm^{-2}. At the eye level of operating personnel or other workers at their regular stations, the leakage did not exceed 4 mW cm^{-2}. At one faulty cleaning door, the leakage was greater than 200 mW cm^{-2}.

Of interest to the reader may be a listing of some specific applications: pharmaceutical drying, opening oysters, film drying, rapid heating of school lunches, pre-cooking onion rings, germination enhancement of tree and alfalfa seeds, drying of paper, vulcanisation of rubber products, veneer drying, drying of leather, pasteurisation of milk, and curing epoxy resin impregnated pipe.

Two other interesting industrial applications of microwaves are the measurement of moisture content in materials, and ZAPPER III® (Oceanography International Corporation, College Station, Texas, USA), a mobile piece of farm equipment which is capable of controlling weeds, grasses, fungi and nematodes in soil. The cost of this latter device is US $250 000; its use eliminates the need for pesticides and herbicides prior to planting.

(c) Microwave ovens The small-size microwave oven used in the home and in some places where food is dispensed by machine, is probably the most notorious potential source of harmful microwave radiation. As a result, the microwave oven has probably elicited the greatest concern of any electronic device with respect to non-ionising radiation.

In 1970, an extensive survey of microwave ovens was conducted to identify those makes and models exhibiting excessive leakage, and to initiate appropriate corrective action.[67] About 4800 ovens, consisting of approximately 42 models produced between 1950 and 1970 by manufacturers of several countries, were surveyed. As a result of the survey, seven manufacturers agreed to implement corrective action programmes to reduce microwave radiation leakage from an estimated 10 000 ovens of the more than 100 000 ovens in use at that time. Ten per cent of those surveyed were found to have leakage greater than 10 mW cm^{-2}. The most frequent fault was due to maladjusted interlocks. The US Federal performance standard for microwave ovens manufactured after October 6, 1972 is 1·0 mW cm^{-2} before sale and

5·0 mW cm^{-2} thereafter, at a distance of 5 cm from any accessible surface.

In the April 1973 issue of the *Consumer Reports* of the Consumers Union of the USA (a publication subscribed to by many of the general public), the following statement was made:

> We believe those extreme tests (shimming the door and operating the oven with very small or no load in the oven) are valid because the conditions they simulate could occur in home use. Gaps can develop in the door seals, either through warping of the door or from wear and dirt build-up. And, since paper towels are used in microwave cooking instead of paper plates and dishes for some items like bacon, they might get caught in the door. Further, it's quite possible for someone to switch the oven on when there's nothing in it. And there's little doubt that people use microwave ovens to heat up tiny amounts of food, particularly left-over snacks—and they wouldn't always be put dead center of the shelf.

Based on these extreme tests, the Consumer Union found leakages in excess of 20 mW cm^{-2} in nearly 50% of the models tested. Furthermore, they recommended to the US Government that the following warnings and instructions be permanently affixed to the front exterior of all microwave ovens:

1. Do NOT operate oven when empty.
2. CAUTION—Microwave radiation may cause pacemaker interference. Persons with pacemaker implants should not be or remain in the same room where this microwave oven is operating.
3. After each use—unplug oven, and check to see that door seal and inside surfaces of door and oven cavity are clean.
4. Do NOT put face close to door window when oven is operating.
5. KEEP OUT OF THE REACH OF CHILDREN. DO NOT permit young children to operate this oven.

The article in *Consumer Reports* brought rapid rebuttals from the microwave oven industry and by some electrical societies, since the Consumer Union had placed a 'Not Recommended' appraisal on microwave ovens. The controversy is not over, but has been kept alive by increasing pressure in the USA to lower the allowable exposure to radiation. This will be discussed later in this book when the exposure standards in eastern European countries (USSR, Czechoslovakia, Poland, *et al.*) are compared with those in the Western World.

In April, 1975, the US Food and Drug Administration ordered notices to be placed on newly manufactured microwave ovens to warn users against radiation dangers. The label reads:

> Precautions for safe use to avoid possible exposure to excessive microwave energy:
> Do not attempt to operate this oven with: (A) object caught in door; (B) door that does not close properly; (C) damaged door, hinge, latch or sealing surface.

A second label will state that the ovens should be serviced only by qualified personnel.

It can probably be stated with accuracy that the electric and magnetic fields radiated by microwave ovens, meeting the required standards, are less than the fields existing around standard electric ranges (*see* Table 2.1).

(d) Communication links The line-of-sight radio relay system, using microwaves, began in 1945 and has mushroomed in the USA and in other countries since then. By 1965, there were 128 000 km of broadband microwave links; by 1969, there were 71 000 microwave towers.[61] Because these towers are very high and because the power levels at which the links operate is low (5 W, typically), their danger to the public is small. However, these towers may be potentially dangerous to the workers who maintain them. Some caution should be exercised at the point of origination of TV and radio programme, such as a sporting event, where a mobile truck will transmit via a microwave beam to a central broadcasting location.

An analytical evaluation has been made by the US Environmental Protection Agency of selected satellite communication systems.[68] Calculations have been made of maximum levels of power density as functions of distance from the source. The radiations of the Goldstone Venus and Mars systems used to communicate with space vehicles are estimated to be 10 mW cm^{-2} at distances of 4·160 km and 9·68 km, respectively. The power densities produced by certain satellite communication systems have been measured. In the worst case, a transmitter of 2 kW produced a power density of 12 mW cm^{-2} at a distance of 150 metres from the antenna. It was concluded that satellite communication systems, operated in accordance with prescribed procedure, should not constitute a thermal-effects hazard, because the radiation is directed upwards. Operated improperly, on-axis radiation power densities may be sufficiently high to create thermally hazardous situations.

(e) Power transmission A number of proposals have been made to transmit substantial amounts of power by means of microwaves. One interesting set of experiments was conducted by the Raytheon Company[69] for the propulsion of a helicopter. Microwave energy was beamed to a small two-metre antenna, which, after being converted to d.c., powered lightweight motors. The system was conceived as a station-keeping device that would hover directly above the transmitter to relay TV or other signals. A more direct method of powering motors by microwave radiation has been proposed by T. K. Ishii.[70]

Today, the most attractive application of microwave power transmission is the synchronous-orbit satellite power station, which converts solar energy to d.c. thence to microwave energy, which is then beamed to a single point on the earth.[71, 72] It is proposed that an equatorial orbit 35 800 km above the earth be used, so that the satellite is exposed to the sun 99% of the time all year round. Power transmitted would be from 3000 to 15 000 MW at a frequency of 2–4 GHz. The proposed total area of the solar panels is 97 km^2. The direct current generated by the solar cells would power microwave crossed-field devices, whose output energy would be transmitted by an antenna 1 km long. The maximum power density impinging on a 43 km^2 multiple antenna system on the earth would be about 870 $W\ m^{-2}$ (87 $mW\ cm^{-2}$). The received microwave energy would be rectified and fed to a common load. The total received power could be carried as direct current by transmission lines or converted and transfomed to high voltage alternating current.

With a power density at the centre of the receiving antenna of 87 $mW\ cm^{-2}$, it would be necessary to locate the receiving antenna system in an isolated location, and for precautions to be taken for workers. Wild life would probably be affected in the immediate vicinity of the antenna system. Control systems would be required to maintain the specified direction of the microwave beam in order to avoid accidental radiation on populated areas.

3

Therapeutic Uses of Electromagnetic Energy

Electricity and magnetism have been used to attempt to alleviate human ailments since early times when they were first discovered. Although it was about 600 B.C. that Thales, a Greek physical philosopher of Miletus, discovered static electricity by rubbing amber with a cloth, it was the shock of the torpedo, an electric fish, that was prescribed by Aetius, a Greek physician, for the treatment of gout, which led to the beginning of electrotherapy. Magnetism was discovered in Magnesia, a part of Asia Minor, in an iron ore known as Magnes stone, magnetite, or lodestone. About 200 B.C., the Greek physician, Galen, is reported to have used a magnet as a purgative.[128] Because live electric fish were not readily available, and because other practical methods of generating electricity had not been invented, magnetism was first used as a therapeutic agent. Two books (in Russian) by Yu. A. Kholodov, *Magnetism in Biology*[73] and *Man in the Magnetic Web (The Magnetic Field and Life)*[74] give an excellent history of the early and present therapeutic uses of magnetism.

3.1 Magnetism

Probably the first notable application of magnetism was made by Mesmer (1774). His experiments were of interest to other scientists, but when he claimed 'animal magnetism' properties for himself by the imposition of his hands, he quickly lost favour with the scientific community and was denounced as a quack and a charlatan.

About 100 years later, a French scientist, Durvel, presented the results of experiments on 100 men using what was called a 'magnetic bracelet'. Two-thirds of these men felt some kind of sensation, or irritation, or itching. Other scientists also reported effects of the magnetic bracelet, so that it was difficult to discount their reports completely.

At a meeting in Moscow in 1966, two papers were presented. L. B. Andreev reported that the first stages of hypertension were relieved by the application of the magnetic bracelet. However, advanced stages were affected only slightly. He also reported that the application of a magnetic field to the back of the head thirty minutes before sleep relieved headaches. N. B. Tjagin reported that symptoms of nervous and vascular diseases disappeared or became less noticeable. In some cases relief was only temporary, but in most cases relief was permanent.

The magnetic bracelet is still used today in parts of eastern Europe, even though there still is no real theoretical basis for its therapeutic properties.

3.2 Electric fields

The controlled use of electric fields and potentials in electrotherapeutic applications began with the invention of various electrical generators and storage devices, such as the spark generator, the Leyden jar, the battery, and electric dynamo in the latter part of the eighteenth century and the early part of the nineteenth century. Most of the early applications involved the direct application of electricity to the human body to obtain muscle and nerve stimulation, and in some cases electroconvulsive therapy.[75, 76] A study of the effects of electricity on the human body (begun about 1930) led to the discovery that fibrillation is the lethal process in death from electric shock. This study led to the development of closed-chest defibrillation (also called cardioversion) and heart massage, procedures which have saved many lives.[129] Electroconvulsive therapy, more commonly known as shock treatment, has proved beneficial to persons suffering from depression and schizophrenia.[130]

Following a long period of little reputable use, electrical stimulation of nerves for the relief of pain is now becoming increasingly popular.[77, 78] This includes the renewed interest of electro-acupuncture, which was introduced into France as early as 1774.[76]

Although the direct application of electricity to the human body is of importance in the field of electrical therapy, it is not of primary interest to the subject of this book. It may be of some interest, however, since its study may reveal various changes which occur in living tissue, which would also occur due to electric and magnetic fields emanating from the sources discussed in Chapter 2.

In the 1780s, experiments were performed to show the effect of living in an electrical field.[76] N. Bertholon believed that the degree of environmental electricity was related to the aetiology of disease. Jean-Paul Marat performed a variety of experiments on himself and his patients in an electrified room to disprove this viewpoint. He found no effects in experiments over a six-year period. This recalls to mind the recent controversy regarding the effects of high electric fields on men working near EHV transmission lines and switching stations (*see* Section 2.2.1 and References 29, 31 and 32).

Reinhold Reiter reports in a review paper[18] in 1964 of a systematic statistical study for the purpose of demonstrating a correlation between changes in reaction times, frequency of diseases, illnesses, accidents, injuries, births, deaths, etc., and changing electric fields, including 'sferics' (electromagnetic radiations in the frequency range of 5–50 kHz), accompanying changing weather. Positive correlations were indicated as follows: diseased, sensitised and healthy persons react to weather changes; a large number of persons are influenced by the weather, completely unknown to them; weather affects individuals ready to respond to it; weather changes may bring about catastrophic responses in persons already in a state of stress.

There is a ramification of the use of electric fields which has apparently proved to be more therapeutic than the mere presence of electric fields. This is the degree and nature of ionisation of various constituents of air.[19, 20] Normally, the number of positive ions exceeds the number of negative ions. When the excess of positive ions is large, there are indications that deleterious biological effects occur in man, such as reduced performance, tenseness and irritability. However, when an excess of negative ions occurs, either naturally, or created by the application of artificial electric fields, beneficial physiological and psychological effects have been reported, such as improved performance, increased work capacity, relief of pain and allergic disorders, more cheerful disposition, and enhanced burn recovery, healing and vitamin metabolism.

Sugiyama,[21] in work on photic stimulation by means of d.c. and a.c. electric fields, found that the critical flicker frequency or CFF (*see* page 5) was increased by an a.c. electric field. It was found that decrease of CFF was a measure of central nervous fatigue. Thus, the application of an electric field could have the equivalent function of minimising fatigue in man.

3.3 Induction Coils

Alternating magnetic fields have been applied to various parts of the body by means of solenoid induction coils.[79] (Electrotherapy of this type is widely used in eastern Europe. See the *Index Medicus*, 'Electrotherapy' for paper citations.) Frequency of excitation was varied from 40–4000 Hz; magnetic field level was about 100 G, which is insufficient to cause visible muscle stimulation or heating. De la Waar and Baker claim that they have detected reflex action by means of a sonic pick-up at locations away from the point of application of the a.c.

magnetic field. They also claim that physiological effects have been obtained, such as, reduction of: blood cholestrol (about 30%), white cell count in blood, erythrocyte sedimentation rate, and possibly blood pressure. They have used this a.c. magnetic field in the treatment of rheumatoid arthritis.

3.4 The Bordeaux Magnetic Machine

Recently there has been interest in what has become to be known as The Great Bordeaux Magnetic Machine Mystery, or 'L'Affaire Priore'. It has been reported in the popular press,[80, 81] and in a number of papers in *Comptes Rendus des Séances de l'Académie des Sciences* since 1964, sponsored by Robert Courrier.

The machine, built at the University of Bordeaux by Antoine Priore, a 62-year-old electrical engineer, was originally used to kill micro-organisms that produce rot in fresh vegetables and fruit. Subsequently, it was reported that wondrously beneficial effects on experimental subjects were obtained by the radiation emanating from the machine. Today, its primary use appears to be in cancer treatment of animals, where a large number of cures have been claimed, as described in further detail below. Of course, human treatment of cancer is the final goal.

The machine is reported to be comprised of the following: a large plasma tube excited by 430 V; a magnetron oscillator (9·4 GHz, 40 kW peak), which is pulsed on for 1 μs at a rate of 1 kHz; two high frequency oscillators (17·6 MHz and 15·8 MHz) and a magnetic field (1200 G) which confines the plasma and which is pulsed at a rate of 50 times per minute.

The magnetron and the two high-frequency signals are mixed in the plasma so that the 9·4 GHz signal is modulated by the two high-frequency signals.

The magnetic field at the table, where the subject is located, is about 600 G. (A second machine was also built which was capable of producing an active magnetic field of about 1200 G.)

Several experiments have been carried out. Some of these are summarised below:

1. A cancerous tissue was grafted to laboratory rats. Twenty-four control rats died within 22–30 days. In those rats which had been exposed to radiation in the magnetic field from Priore's machine, the cancerous tissue was either completely absorbed, or, if the treatment was begun after cancer had started to spread, had regressed to the point of total cure. There were no observed bad side effects and there was no recurrence of the cancer.[82]

2. A different and highly malignant strain of cancerous tissue, leukaemia, which can cause death in as little as two weeks, was grafted onto laboratory rats. Again, exposure to the radiation in the magnetic field resulted in the complete inhibition of growth of the cancerous grafts, or, if the disease had started to develop before the treatment was begun, to its total regression.[83]

3. Mice were inoculated with a microscopic blood parasite (*Trypanosoma equiperdum*) which causes sleeping sickness and death in a short time. The course of the experiment could be followed closely by analysing blood samples taken daily. All control animals died by the fifth day. Mice exposed daily to the radiation and magnetic field (starting one hour after inoculation) and continuing 10–15 days usually survived (38 out of 46 animals) with complete disappearance of the parasites from their systems. They also developed a specific immunity to further infection. Animals exposed some days before inoculation appeared to have developed an immunity.[84]

4. Rabbits, which develop a chronic illness when inoculated with trypanosomes from which they usually die after several weeks, were tested with the radiation in a magnetic field with the same positive results as described above.[85]

A twenty-man commission composed of well known men of science, medicine, law and other professions, have begun and are continuing investigations into the claims made for this mystery machine. They have been able to report experimental findings, but have not been able to provide an explanation of the workings of the Priore equipment.

Funding has been provided by the French Government to construct a third and more powerful machine. Time will tell whether this has been a great hoax, whether the French Government and numerous well known scientists have supported a charlatan, or whether mankind has stumbled on a revolutionary scientific development.

3.5 Electro-diathermy

Electro-diathermy is probably the most acceptable form of application of electric and magnetic fields to man for therapeutic purposes. The first method utilised the capacitative currents flowing in the human body when the body is placed between two large electrodes. Later, an electromagnetic field created by a coil was used to produce eddy currents in tissue. For example, a limb was placed inside a solenoid-type coil, or a pancake coil was placed on the part of the body in need of heat therapy. The frequency of early diathermy machines was about 12 MHz.[75] Today, 27 MHz is used in the USA, primarily because of the allocation by the Federal Communication Commission (FCC) of this frequency band for industrial use. Microwave diathermy provides more localised heating because of the shorter wavelength and the ability to beam the electromagnetic energy by radiators, such as horns. For example, in human thighs, a uniform temperature rise to 40·1°C five centimetres deep can be obtained.

Eddy currents are greater in tissue of greater conductivity. In man, the tissues in decreasing order of conductivity are muscle, brain, fat, skin, and bone.

The primary effect of electro-diathermy is physical, that is, there is a temperature rise. However, a secondary effect occurs which is physiological; hence, careful technique and carefully measured dosages are required.

A study was made in 1970 of diathermy treatments in the St Petersburg, Florida (Pinellas County) area.[86] It is estimated that, per month, 13 000 individuals received 90 000 diathermy treatments, consisting of 25 000 short-wave, 20 000 microwave, and 45 000 ultrasonic treatments. They were applied by largely unskilled technicians with little or no knowledge of the amount of the power or energy actually being dissipated in the persons being treated. Levels of exposure were substantially above those normally considered 'safe'.

Clinical effects of diathermy which have been claimed are as follows:[75]

1. On circulation
   a. Local effects, such as, active arterial hyperaemia (dilatation of capillaries), increased lymph flow, and marked increase of secretion in glandular organs.
   b. General effects, such as, rise of body temperature, increased pulse rate, respiration rate, and general body metabolism.
2. On the nervous system, marked sedative effect on irritative conditions of sensory nerves (pain) and motor nerves (spasms and cramps).

Some clinical uses of diathermy are as follows:

1. Sub-acute and chronic inflammatory and congestive conditions and circulatory disorders.
2. Traumatic and inflammatory conditions of bursae, bones and joints after the acute stage.
3. Spastic conditions of the stomach, gall-bladder, and intestines.
4. Acute, sub-acute and chronic forms of bronchitis and pleurisy, and relaxation of spasms of bronchial asthma.
5. Neuritis and neuralgia.

There are dangers of excessive dosages which could cause burns, particularly of the eyes. Diathermy should not be used in cases of acute inflammatory processes accompanied by fever and infection, on malignant tumours, or in cases where there is a tendency to haemorrhage.

There is an interesting application of microwave diathermy which bears mention. Microwave heating of the uterine wall during parturition at child-birth, at the time of contraction, has been found to have an analgesic effect. Births treated in this manner progressed more smoothly than without the diathermy treatment.[87]

3.6 Athermal Therapy

Presman[88] describes a diathermy which is 'athermal', that is, there is no discernible temperature rise. He speculates that low non-thermal intensities might be more effective than thermal intensities. Areas of most promising use, so far, are for symptomatic treatment of hypertonia of neural origin, effect on cardiovascular regulation, stimulation of leucocyte production in connection with radiation injury, and inhibitory action on malignant tumours.

Athermal diathermy has also been used in the form of narrow radio frequency pulses of very high intensity.[75] There have been many claims of satisfied patients from such treatments. However, the majority opinion of scientists and physicians is disbelief; they claim there still needs to be substantial investigation before athermal diathermy becomes an accepted type of treatment.

3.7 Discussion

The controversy over athermal diathermy is of great importance to the final judgment regarding the significance of electrical pollution. The reader can appreciate the possibility that if athermal diathermy is proved to be an efficacious method of treatment, then other low-level methods of electrical treatment may also be useful. By the same token, electrical pollution, which is an uncontrolled application of electric and magnetic fields, would be expected to have some physiological effect on man.

Probably the leading advocate in the USA of the effect of low-level electric and magnetic fields is Robert O. Becker, M.D., of the Veterans Administration Hospital, Syracuse, New York. His work[89] and the work of others, has gained a certain degree of legitimacy. The New York Academy of Sciences sponsored a conference on electrically mediated growth mechanisms in living systems; it was favourably reported on in the *Journal of the American Medical Association.*[90, 91] Dr Becker's remarks at the N.A.S. Conference are summarised as follows:

> The true depth of knowledge in this area is extremely shallow. The significance of all this is that electromagnetic phenomena do have biological effects. We know that. We also know that these effects can be good or bad and that it seems evident that solid-state electrical properties are present. In addition to data sending involving the central nervous system, there may be another, concurrent system located within the cell, such that each cell contains its own data transmission system that does not depend on commands from a central point, but may react independently to stimuli of an electric or chemical nature.

He also stated[89] that three life processes in mammals have been changed by electromagnetic fields, which can probably occur in man. These are: stimulation of bone growth; stimulation of partial multi-tissue regenerative growth; and, influence on the basic level of nerve activity and function.

Dr C. Andrew L. Bassett, Professor of orthopaedic surgery at Columbia University College of Physicians and Surgeons, has recently reported the successful utilisation of pulsed low-frequency, low-intensity electromagnetic fields to promote bone regeneration in five children with congenital pseudarthosis.[92, 93] Electric fields in the bone are generated by coils on each side of the bone fracture. Two types of stimulation have been used: frequency of 1 Hz, field in bone of 2 mV cm^{-1}; and 65 Hz, 20 mV cm^{-1}. The power level in the bone is about 0·1 mW cm^{-2}. Treatment time is about one-half that of conventional methods.

4

A Study of Various Occurrences of Known Electrical Pollution

At the outset, one is forced to say that there have probably been relatively few recognised occurrences of known electrical pollution. One can speculate that if such occurrences were widespread, there would be greater public interest in electrical pollution. As it is, the general public is oblivious to the possibility, or perhaps, the probability, that they are being affected adversely by the electricity and magnetism around them. Perhaps this book, and particularly this chapter, may awaken more people to this possibility.

We must distinguish between effects on man which are 'thermal' as opposed to those which are 'athermal'. There are, of course, more known cases of 'thermal' effects, since they are recognisable.

Thermal effects on man have generally been accidental; these are often reported when they are of occupational origin. Athermal effects are almost impossible to detect. Some studies have been made with human volunteers, subjected to controlled electrical pollution, to determine if there really are such effects. A technique which can be utilised to determine possible effects of electromagentic fields is the *epidemiologic survey*. In these surveys, computer analyses of large groups of individuals under various types of exposure can be made as reflected by clinical symptoms, electro-physiological and biochemical testing. Statistical analyses of these groups and appropriate control groups could yield significant results. The results of several elementary surveys have been incorporated in what follows.

4.1 'Athermal' Electrical Pollution

There are only a few known cases of electrical pollution which can probably be classified as 'athermal'. These will be discussed first. Included in the discussion are thc results of studies on human volunteers.

4.1.1. Phosphenes

The phenomenon of phosphenes is the sensation of light flashes in the eye. The flashes are colourless or faintly tinted blue or yellow. They may be produced by an alternating magnetic field of about 800 G. Frequency of the stimulus, which is applied to the head, is 30–40 Hz for peak response.

Michael D'Arsonval was probably the first to notice magnetic phosphenes in 1893. Harold S. Alexander[94] gives a history of work done on phosphenes up to 1962. This survey, which covers most of the tests and studies to date, suggests that the cause of phosphenes may be in the retina of the eye, presumably being the result of eddy currents. Pressure on the eyeball can stop the visual sensation temporarily. With constant stimulus, a decrease in the brightness of the phosphene occurs with time, probably due to fatigue.

The tests described above were performed on volunteers, presumably with no harmful after-effects. However, occurrences of phosphenes have been reported in workmen in a nitrate factory at Notodden, Norway, probably from large choke coils used to limit the current supplied to electric furnaces.[95]

I have heard of an experiment which was performed which might be considered an extension of the experiments with phosphenes. In addition to the low-frequency magnetic field, a d.c. field was also applied to the head, together with ultrasonic excitation through an electrostatic speaker excited by a piezoelectric transducer at a radio frequency. The ultrasound was coupled to the top of the head by a water-filled chamber. Human volunteers reported small muscle quivers in the eyelids, tingling in the hands, and some dizziness. The tests were terminated soon after they were started because of possible unknown effects on the brain which might be permanent, even though calculations indicated that the induced currents were too low to be dangerous.

4.1.2 Radiosounds

The principal researcher in the study of radiosounds was Allan H. Frey.[96, 97, 98] He has demonstrated by his experiments that people can 'hear' electromagnetic energy within the frequency range 200 MHz to 3000 MHz when this electromagnetic energy is modulated in some way. The sound is of the nature of buzzes and hisses, and is probably detected in the temporal lobe of the brain, appearing to be heard within or immediately behind the head. The threshold of perception of this radiosound is dependent primarily on peak power density, and not on average power density. For example, for four tests at 425 MHz, the peak power density threshold varied only from 229 to 271 mW cm^{-2}, whereas the average power density varied seven-fold from 1·0 to 7·1 mW cm^{-2}. The threshold level increases rapidly above 1500 MHz due to the fact that less energy is able to penetrate the head because of a shortened skin depth. Below 300 MHz the body becomes more and more transparent, with no appreciable absorption. In the frequency range of interest, approximately 40% of the incident energy is absorbed. Perceived characteristics of pitch and timbre appear to be functions of modulation.

There have also been reports of some people hearing aurora displays and meteors entering the earth's atmosphere. Certain individuals are extremely sensitive to radiosounds.

4.1.3 Effect on the Cardio-vascular System

Scientists in the USSR have collected a considerable amount of data from clinical investigations of persons who have been chronically exposed to electromagnetic radiation. Presman[99] gives a good summary of this work. Effects on the cardiovascular system are listed in three categories, as follows: reduced blood pressure; slow heart beat; electrocardiograph (ECG) changes.

The frequency range covered in the studies extended from about 300–10 000 MHz. The effects were most pronounced at the high end of this frequency range for which electromagnetic energy is absorbed in the surface tissue of the human body. Power levels at UHF and microwave frequencies were about 1 mW cm^{-2}; at short-wave and medium-wave frequencies, the field strengths were 10–100 V m^{-1} and 100–1000 V m^{-1}, respectively. It is postulated that these changes in the cardiovascular system are due to the direct action of electromagnetic energy on the surface receptors of the nervous regulatory system.

Michaelson[100] reviewed and analysed presentations given at the International Symposium on Biologic Effects and Health Hazards of Microwave Radiation, held in Warsaw, 15–18 October, 1973. He reported that clinical syndromes associated with microwave exposure among several occupational groups were presented which included some parameters of cardiovascular and haemodynamic disturbances.

4.1.4 Other Effects

Various other effects have been reported by scientists in the USSR.[99]

(a) A conditioned vascular reflex has been obtained upon the application of a high-frequency field with an intensity of 10 000 V m^{-1}.[101]

(b) People who have worked for a long time in microwave fields have a higher amplitude of slow wave in electroencephalogram (EEG).

(c) There was a reduction in the sensitivity of the olfactory mechanism in people working with generators operating between the short-wave and microwave frequency bands.

(d) There are changes in blood indices in people who have been exposed for a long time (more than one year) to weak electromagnetic fields over a wide variety of frequencies. Some changes have been: percentage of haemoglobin, protein composition and content, and histamine content.
(e) There was a disturbance in carbohydrate metabolism to chronic exposure to low-intensity microwave fields. The sugar content of the blood and urine was increased, and the sugar curves had a pre-diabetic form.

4.1.5 *Effects of Geophysical Parameters*

The correlation of biological parameters (blood pressure and leucocyte count, among others), illnesses, and deaths, with geophysical disturbances has been the subject of interested study since 1935. The experimental investigations suggest that the main role is the change in environmental magnetic and electric fields and atmospherics. Brief reviews of these studies point out the following cases:[12, 13, 102]

(a) Incidence of cerebrospinal meningitis in New York, and incidence of relapsing fever in the USSR was correlated with solar activity.
(b) Rate of psychiatric hospital admissions was correlated positively with geomagnetic storms.
(c) Total number of admissions to hospitals and number of deaths in hospitals was correlated with geomagnetic storms.
(d) Mortality from nervous and cardiovascular diseases was correlated with magnetic storms in Copenhagen and Frankfurt-on-Main.

Porkorny and Mefferd[103] examined the relationship between 2017 homicides, 2497 suicides and 4953 psychiatric hospital admissions and seven related measures of geomagnetism. The study involved 1096 days in the years 1959–61, which happened to be a period of decreasing solar activity. They concluded that geomagnetic fluctuations do not influence psychiatric hospital admissions, suicides, or homicides.

In order to clarify these claims and the controversies accompanying them, Friedman, Becker and Bachman[104] performed controlled experiments on volunteers exposed to artificially generated magnetic fields. Simple reaction-time performance was examined in both schizophrenic and normal individuals. No changes could be detected to steady fields of 5 G and 17 G. However, when these fields were modulated at rates of 0·1 Hz and 0·2 Hz, definite, statistically significant, temporary changes in reaction time were observed. Further experiments with humans were discontinued when Kholodov reported observations of cerebral gliosis in subjects exposed to these fields.

It can be conjectured that if geomagnetic fluctuations do have the effects itemised above, then the near-absence of such fields should produce detectable changes in some physiological or psychological parameters. Beischer[14, 15] has performed such near-zero magnetic field experiments. The only change he observed was a decrease of the critical flicker fusion threshold (*see* page 5).

4.1.6 *Effect of High-Level Power Frequency Electric Fields*

The Russians have published a number of papers stating that high-level electric fields due to extra-high-voltage (EHV) transmission lines and switching stations caused disorders of the functional state of the nervous and cardio-vascular system.[31, 37] They describe a number of complaints such as listlessness, excitability, headache, drowsiness and fatigue. Their concern has led to the establishment, by the USSR, of standards and regulations for the protection of labour during work on substations and transmission lines at voltages of 400 kV, 500 kV, and 750 kV.[32] These standards allow unlimited exposure for electric field intensities below 5000 V m^{-1}. However, above 5000 V m^{-1} exposure time is limited. On a 24 hour basis, the limits are: 180 minutes at 10 kV m^{-1}, 90 minutes at 15 kV m^{-1}, 10 minutes at 20 kV m^{-1}, and 5 minutes at 25 kV m^{-1}.

In the USA, a study which extended over a period of nine years, claimed that no adverse effects were observed. A team of ten linemen, some of whom worked on 765 kV lines, were carefully examined and followed up by a team of physicians at The Johns Hopkins University, Baltimore, Maryland. Each lineman was examined seven times during the nine-year period, including a session, each time, with a psychiatrist. There were no significant changes of any kind, physically or emotionally that could be related to their work.

The US Environmental Protection Agency has begun a study of the health and environmental effects that may be involved in the vicinity of 765 kV transmission lines.

An interesting anomaly has been observed in medical tests of signal transmission along nerve fibres. The test involves stimulation of peripheral nerves and subsequent detection of the response on the surface, or slightly below the surface, of the scalp. The test is known as the *evoked-potential response*. When stimulation is of nerves in the legs or arms, the test is called *somatosensory*; when stimulation is by means of light, it is *visual*; when by means of sound, it is *auditory*. In specific studies by the Department of Neurology, School of Medicine, University of California, Los Angeles, California,[105, 106] the stimulation frequency was made variable, and the average peak-to-peak amplitude of the response was measured. It has been observed that in *normal* subjects, the curve of amplitude as a function of frequency shows a substantial increase at 60 Hz. A variety of experiments were performed attempting to attribute this to an artifact caused by the presence of the power line frequency in the equipment. However, no significant amount of such 60 Hz activity was detected. Could it be that the human body, being exposed at all times to the power line frequency, has developed a sympathetic response at the power line frequency? If so, the peak in the evoked response should occur at 50 Hz in European countries.

It is interesting to note that the heart is most susceptible to the danger of ventricular fibrillation, due to electrical shock, for a frequency range of 40–80 Hz.[107]

Finally, tests made in Japan by Sugiyama demonstrate that the critical flicker frequency was affected by a.c. and d.c. electric fields. The electric fields were generated by electrodes above and below the human subject, with a voltage of 15 kV.[21]

4.2 'Thermal' Electrical Pollution

The primary sources of 'thermal' electrical pollution occur at radio frequencies, especially at ultra-high frequencies and microwaves. It can be correctly stated that more physical harm to humans was experienced in the past than is experienced today. The 'Radiation Incidents Registry Report, 1970', published by the US Bureau of Radiological Health[51] reports two deaths claimed to be caused by excessive microwave radiation. The first case involved a 28-year-old female assembler in a microwave equipment and manufacturing company in California. She was reported exposed to radiation from a radar unit for a period of six months during 1952 and 1953. She haemorrhaged and died. The second case involved a 42-year-old electronics technician at a microwave equipment manufacturing company in California. He reportedly worked ten feet from a radar transmitter. He experienced tissue destruction and died in 1954.

Today, governments, the armed forces, industries, scientists, engineers, and consumer groups are more aware of the possible dangers of electromagnetic radiation. This awareness has lessened, but has not eliminated, the occurrences of physiological damage to humans. Damage usually occurs unbeknown to the victim.

There is a current trend which has hidden electromagnetic radiation accidents. Although it was written about shock accidents in hospitals and clinics, what John M. R. Bruner says in a short report,[108] 'The Horrors of Common Practice', seems to apply.

> There is a lack of useful information in an era infatuated with communication. Accident reports have virtually disappeared from journals; chilly responses meet inquiries directed to agencies whose business it should be to know about electric injury. Without reliable data on the incidence and cause of illness, intelligent planning of preventive measures is not possible.

The current situation can probably be summarised by a viewpoint of a businessman/inventor which appeared in an industrially-oriented publication, *Electronic Engineering Times*, and an opposing letter by a design engineer in a subsequent issue.[109] The proponent stated:

> For over ten years, I have deliberately exposed myself and others to direct, open door exposure to microwaves emerging from operating 1- and 2-kW microwave ovens. I found it a pleasant experience with no ill effects.
>
> Years ago while fixing microwave ovens with their doors wide open and their safety switches inoperative, I was frequently bathed in microwave radiation. It was a most pleasant, warm, and comfortable feeling. As doctors were advocating and using microwaves to make sick people well, I had no fear.

The opponent retaliated:

> As a result of accidental exposure to S-band radar (3 GHz), I lost the central vision in one eye, and both eyes are scarred on the retina. There was no associated pleasant, warm and comfortable feeling. There was, in fact, no knowledge or awareness at the time of exposure. My case is not isolated. Two co-workers were victims of prior accidental exposure, with resultant retinal burns and partial blindness.

A review of the literature regarding the study of the effects of microwaves on man will probably *not* resolve the opposing viewpoints expressed above, since the literature presents conflicting conclusions.

4.2.1 Microwave Cataracts

The foremost effect of microwave radiation which has been recognised is the formation of cataracts on the lens of the eye. A critical review of this subject has been made by Milroy and Michaelson.[110] Table 4.1 summarises the studies of this review. A chronological order has been attempted.

In experiments with animals, cataracts are formed both on the anterior and on the posterior locations of the lenses of the eyes. Anterior cataracts are associated with gross eye damage, whereas posterior cataracts are delayed and occur at lower levels of microwave exposure. Frequencies of between 1000 and 3000 MHz are the most hazardous. Below 1000 Mhz, the doses of exposure either do not produce cataracts, or if they do, they are lethal. Above 3000 MHz, cataracts are usually anterior and are thermally generated.

The work of Cleary and Pasternack (No. 11 in Table 4.1) suggested that microwave exposure has an ageing effect on the lens. Zaret (No. 15 in Table 4.1 and see Reference 113) described the pathological progression of microwave cataracts. It begins as roughening, thickening, and opacification of the posterior capsule. It is believed that this affects metabolic exchange across the capsule resulting in asphyxiation and loss of transparency of the lens. Zaret set the threshold for human microwave cataracts at 100 mW cm^{-2}.

With regard to two other questions, Milroy and Michaelson[110] state that additional experimental work is required to determine: (1) if there is a cumulative sub-threshold effect, and (2) if there is a non-thermal effect. Appleton[112] is quite adamant in his comments. In his study (No. 17 of Table 4.1), he states, 'It is reasonable to conclude that no lens damage was caused by occupational exposure to microwave energy.' He doubts that cumulative effects occur and thinks that lens damage could not occur in a human from acute exposure without associated severe facial burns. He believes that microwaves are safe to humans if existing standards of safety are observed.

4.1.2 Thyroid Effects

Scientists in the Soviet Union have reported thyroid changes in workers exposed to microwaves. They noted enlargement of the gland in some cases, along with an increased absorption of radioactive iodine by the thyroid. However, there were no clinical symptoms of toxic diffuse goitre. They suggested that these effects may be due to a direct neurological effect resulting in stimulation of the hypophysis (the pituitary gland located at the base of the brain), which caused an increase in the hormone, thyrotropin.

Milroy and Michaelson surmise that the reported effects in humans may be due to thermal stress and homeostatic adaptation to maintain normal body temperature. Their work on rats[114] does not appear to support the concept of a direct 'athermal' effect of microwave radiation on the neuroendocrine system. They suggested that further studies of the neuroendocrine responses be undertaken in higher animals.

TABLE 4.1

Review of eye disorders due to microwave radiation

No. and Year	*Case*	*Exposure*	*Biological effects*	*Researchers*
1 (1943)	Group of 45 military radar operators	Incidental to manufacture and testing of microwave generators, and operation and repair of radar units	Did not note any eye changes	Daily[131]
2 (1952)	20-year-old radar repair man	—	Bilateral cataracts	—
3 (1952)	32-year-old microwave generator operator	4000–5000 MHz 500 W (peak) 250 W (average) 40–380 mW cm^{-2} and up to as much as 1160 mW cm^{-2}	Bilateral posterior subcapsular opacities. Left eye: cells in the aqueous and vitreous humours, vitreous opacities, and choroiditis. Lens removed. Right eye: cataract in right lens.	Hirsch and Parker[132] Hirsch re-evaluated in 1970[111]
4 (1958)	335 microwave workers	400–9000 MHz 1 MW peak radiated power	Found no ocular pathology	Barron and Baroff[133]
5 (1959)	22-year-old technician	2500–3000 MHz over a two-month period	Anterior and posterior cataracts bilaterally.	Shimkovich and Shilvaev[134]
6 (1961)	Microwave worker	High intensities	Haziness in the posterior lens suture.	Minecki[135]
7 (1961)	475 exposed workers 359 control workers	At military and civilian installations (several hundreds mW cm^{-2})	Found only a slight but statistically significant difference in posterior polar changes	Zaret *et al.*[136]

8 (1962)	370 persons working with microwave generators (two case studies are included)	—	Opacifications described as points, threads, and dust, were noted in the anterior-posterior cortical layer close to the equator	Belova[137]
9 (1965)	Retrospective study in radar workers using US Veterans Administration Hospital records. 2946 with cataracts compared with 2164 without cataracts	Military radar	No increased risk of cataracts was found	Cleary, Pasternack and Beebe[138]
10 (1965)	Two cases	—	Mention microwave cataractogenesis. First case: no ocular pathology was noted. Second case: lenticular opacities observed	52 Abstracts covering Soviet literature from 1937 to 1965.
11 (1966)	Statistical analysis of 736 microwave workers and 559 controls	—	Increased rate of accumulation of posterior polar defects was noted in microwave workers; correlated to the duration and degree of exposure. Number of defects increased with age in both groups.	Cleary and Pasternack[139]
12 (1968)	51-year-old technical writer	Incidental exposure for 7 years to microwave radiation of a variety of wavelengths	Eight years after last exposure, underwent lens extraction	Kurz and Einaugler[140]

TABLE 4.1—*continued*

No. and Year	*Case*	*Exposure*	*Biological effects*	*Researchers*
13 (1968)	200 microwave exposed personnel 200 controls	600–10 700 MHz. Power within limits acknowledged by regulations	No clinically significant lenticular changes, but a statistically significant increase in 'punctate and nebulous opacities' were observed. Both anterior and nuclear as well as posterior cortical opacities are reported	Majewska[141]
14 (1968 1969)	Microwave workers	—	Significant lenticular opacities	—
15 (1968 1969)	—	—	In various prior studies, found 44 known occurrences of microwave cataracts (includes the two cases of No. 3 and No. 5, above)	Zare[142, 143]
16 (1970)	US *Radiation Incidents Registry Report 1970*	Microwave radar and generators	Occupational: 3 cases of ocular irritation 3 cases of cataracts Non-occupational: 2 cases of ocular irritations or burns	Mills and Segal[51]

			Workmen's Compensation Claims: 6 cases of lens opacities 3 cases of eye trouble 3 cases of cataracts (Most claims were denied or no action was taken)	
17 (1974)	Five year study of workers exposed to microwaves and controls	Military environment	No lens damage was caused by occupational exposure to microwaves. In 7 cases of presumed acute exposure, no physical evidence of superficial burn production, no evidence of eye damage	Appleton[112]
18 (1974)	19 members of staff of Physical Medicine Service at Walter Reed Army Medical Center, and controls	Microwave diathermy equipment	There appeared to be no difference between workers and controls	Appleton[112]
19	Potato farmer	Lived next to Grumman Aerospace in Calverton, New York (less than 10 mW cm^{-2})	Cataracts	Investigated by IEEE Committee on Man and Radiation (COMAR)

4.2.3 Other Effects

The *Radiation Incidents Registry Report 1970* of the US Bureau of Radiological Health[51] has reported a number of other effects of exposure to high frequency and microwave radiation besides cataracts, eye burns and lens opacities. These are: skin lesions and burns; genital complaints; blood disorders, including leukaemia; stress syndrome; temporary adrenal insufficiency; fainting and fatigue.

5

Regulatory Definitions of Low-level Electric and Magnetic Fields

A convenient method of specifying what is meant by low-level electric and magnetic fields is to examine human exposure standards. Exposure standards have not been static, but have changed with time. Safety levels have usually been lowered as knowledge of the effects of electric and magnetic fields has increased. Furthermore, exposure standards are not uniform throughout the world; safety levels are generally significantly lower in eastern European countries compared with western Europe and the USA. The following listing gives accepted or proposed exposure standards for various types of radiation in several countries.[115, 116]

5.1 USA

5.1.1 Microwaves (300–30 000 MHz)

(a) The USA Standards Institute, in 1966, formulated a Standard (C-95.1), entitled 'Safety Level of Electromagnetic Radiation with Respect to Personnel', which sets the protection guide at 10 mW cm^{-2} as averaged over any possible 0·1 hour period. If the exposure time is less than 0·1 hour, an energy density of 1 mWh cm^{-2} (milliwatt-hour per square centimetre) is used to limit the allowable radiation power density. Unlimited exposure time is allowed at a power density of 10 mW cm^{-2}. For power densities in excess of 10 mW cm^{-2}, the allowable exposure time (ET) per 0·1 hour is determined by the formula,

$$\mathrm{ET} = \frac{1\ \mathrm{mWh\ cm^{-2}}}{P\ \mathrm{mW\ cm^{-2}}}\quad \text{(in hours)}$$

For example, if *P* is 60 mW cm^{-2}, the allowable exposure time is 1/60 hour or 1 minute per 0·1 hour (that is, 1 minute per 6 minutes). In addition, the USASI Standard states the following:

> Under conditions of moderate to severe heat stress, the guide number given should be appropriately reduced.

Under conditions of intense cold, higher guide numbers may also be appropriate after careful consideration is given to the individual situation.

The Standard indicates that exposure levels of 1 mW cm^{-2} will not result in any noticeable effect on mankind. This Standard was reaffirmed with only minor modifications in 1974.

(b) Standards have been set for microwave ovens by the Federal Bureau of Radiological Health. The radiation level at a distance of 5 cm from the external surface of the oven shall not exceed 1 mW cm^{-2} when the oven is new, and shall not exceed 5 mW cm^{-2} within the life-time of the oven. Continuous exposure is assumed; frequency is most frequently 2450 MHz.

5.1.2 *Radio frequencies*

For radio frequencies in the range 30 kHz to 30 MHz, the protection guide for continuous whole body radiation is 0·3–1 mA cm^{-2}, or 100 V m^{-1}.

5.1.3 *Low frequency, particularly 60 Hz*

In the hospital environment, one of the most important considerations is the danger of ventricular fibrillation. Several different organisations, Underwriter's Laboratories, the Association for the Advancement of Medical Instrumentation, the National Fire Protection Association, and the US Standards Institute C101 Committee, are all concerned with the hazards of electrical equipment. Of particular importance is the setting down of safety limits of current which the body can pass without physiological damage, not so much from the direct application of power line voltage, but from leakage currents in the ground circuits which may occur due to inadequate earthing (grounding) of electrical equipment, or due to faults which may occur in the equipment.

For equipment (Type A) which is used with electrically susceptible patients, the maximum leakage current, for wires connected to the patient, is 10 μA (RMS—Root Mean Square) or 14 μA (d.c.), and for contact with the enclosure or chassis, is 100 μA (RMS) or 140 μA (d.c.).

For equipment (Type B) which is usually used where contact with electrically susceptible patients is unlikely, the maximum leakage currents for wires connected to the patient is 50 μA (RMS) or 70 μA (d.c.), and for contact with the enclosure or chassis, is 500 μA (RMS) or 700 μA (d.c.).

These standards apply for frequencies up to 1 kHz. Above 1 kHz, the leakage currents can be larger by a multiplicative factor equal to the frequency in kHz up to a current of 2 mA for Type A equipment and up to 10 mA for Type B equipment.

The generally accepted limit of electric field intensity at 60 Hz is 15 kV m^{-1}.

5.2 USSR

5.2.1 Microwaves (300–300 000 MHz)

The following are maximum permissible intensities of microwaves above 300 MHz for the time durations listed.

10 μW cm^{-2}—working day
100 μW cm^{-2}—2 hours daily
1000 μW cm^{-2}—15 minutes daily

5.2.2 Radio Frequencies (100–300 MHz)

Exposure standards for radio frequencies are as follows:

0·1–3 MHz—20 Vm^{-1}
3–30 MHz—5 V m^{-1}
30–300 MHz—5 V m^{-1}

The installation of new radio frequency generators, whose power outputs are greater than 40 kW, requires a separate room whose area must be greater than 25 m^2.

5.2.3 Power-Line Frequency (50 Hz)[32]

Unlimited exposure is allowed for field strengths below 5000 V m^{-1}. However, for larger field strengths, the following are permissible times of exposure:

10 kV m^{-1}—180 minutes
15 kV m^{-1}— 90 minutes
20 kV m^{-1}— 10 minutes
25 kV m^{-1}— 5 minutes

5.3 West Germany

The critical limit of microwave radiation intensity for human exposure is 10 mW cm^{-2}. No allowance is made for time of exposure.

5.4 Netherlands

The N. V. Philips Company in Eindhoven has established extensive criteria for the protection of its employees. For microwave radiation (30–300 000 MHz), the regulation states,

> Radiation intensities higher than 10 mW cm^{-2} should be considered dangerous. Safety precautions should, however, be based on a permissible level of 1 mWh cm^{-2} (milliwatt-hour per square centimetre).

These levels are identical to those specified in the USA by the C-95.1 Standard of the American National Standards Institute.

5.5 Poland

For microwave radiation (300–300 000 MHz, the maximum allowable mean values of power intensity are as follows.

1000 μW cm^{-2}—only in case of emergency and with special protective measures
100–1000 μW cm^{-2}—20 minute maximum per day
10–100 μW cm^{-2}—2 hours maximum per day
10 μW cm^{-2}—no limitation

5.6 Czechoslovakia

This country probably has the broadest set of regulations for microwave and radio frequency radiation of any country in the world.[117] For those who work near or operate high frequency and ultra-high frequency generators, the following are allowable intensities for an eight-hour period:

30 kHz–30 MHz—50 V m^{-1}
30–300 MHz—10 V m^{-1}
300 MHz–300 GHz (CW)—25 μW cm^{-2}
300 MHz–300 GHz (pulsed)—10 μW cm^{-2}

The lower average power of pulsed radiation is based on the findings that pulsed microwaves are biologically much more effective than continuous wave (CW). This is discussed in greater detail in the next chapter.

For those not working near HF and UHF generators, and for the general population, the following are allowable intensities for a 24-hour period:

30 kHz–30 MHz—5 V m^{-1}
30–300 MHz—1 V m^{-1}
300 MHz–300 GHz (CW)—2·5 μW cm^{-2}
300 MHz–300 GHz (pulsed)—1 μW cm^{-2}

5.7 United Kingdom

In standards recommended by British officials, the upper permissible limit of microwave radiation (30–30 000 MHz) is 10 mW cm^{-2} (averaged for pulsed operation), with no reference to time of exposure.

5.8 France

The French military exposure criterion is 10 mW cm^{-2} or one hour or longer. For larger exposures, the allowable time of exposure (T_p) is given by the formula,

$$T_p = \frac{6000}{P^2} \text{ (in minutes)}$$

down to values of T_p of 2 minutes. For example, for $P = 20$ mW cm^{-2}, the allowable exposure time is 15 minutes. This formula is the same one which was adopted by the US Army and US Air Force in 1964 and 1965. A 55 mW cm^{-2} limit is set by the above formula due to the difficulties in controlling exposures of less than 2 minutes.

5.9 Discussion

Obvious in the above listing is the difference of allowable exposures between the USA and western Europe and the USSR and eastern Europe for UHF and microwave radiation. The USSR specifies 10 μW cm^{-2} for an eight-hour working day; Poland allows unlimited exposure at this level; Czechoslovakia allows workers 25 μW cm^{-2} for an eight-hour day, but only 2·5 μW cm^{-2} for unlimited exposure for the general public. The USA allows an unlimited exposure time to 10 mW cm^{-2}—one thousand times greater than that of the USSR. This fact has caused an uneasiness in the minds of some scientists and engineers in the USA, that perhaps they have overlooked something. Consumer-type groups are even more suspicious. Most experimentation in the USA has been short term. Lacking are studies extending over long durations, such as ten years (except the study by Singewald, *et al.*[29] at Johns Hopkins University, which covered a period of nine years). It is believed that no firm conclusions can be made without the results of such long term studies.[118]

It should be pointed out that the Soviet studies which led to the 10 μW cm^{-2} radiation exposure standard dealt with effects on the central nervous system, and included behaviour in humans and pathology in animals. The USA studies, on the other hand, were concerned primarily with more observable physiological changes in humans and animals. The USA standards set levels of exposure to avoid 'thermal effects', whereas the USSR standards set levels of exposure to avoid 'athermal effects'. Assuming that both are correct, we can say that 'athermal effects' occur at exposure levels between the two standards.

Another method of specifying what is meant by low-level electric and magnetic fields is to determine the levels of currents, current densities and fields which elicit sensation in man.

The most hazardous band of frequencies is centred around 50–60 Hz. Three responses are of importance, namely: perception of electric current flow; uncontrollable muscular contraction, especially of cardiac muscle, and death. Owing to physiological variations in humans, threshold levels for the above responses are usually defined in terms of a certain percentage of a population.

We are concerned with the smallest level, namely the perception of electric current. Keesey *et al.*[27] state that the 1% perception threshold is 0·13 mA, and the 50% perception threshold is 0·35 mA. Dalziel[119] presents a graph of percentile rank as a function of perception current for men; his 1% and 50% levels are 0·5 mA and 1·1 mA respectively. European and USA standards organisations have adopted 0·5 mA as the maximum permissible leakage current from appliances. This current is for external contact with intact skin, such as from hand-to-hand or hand-to-foot. The total resistance between points of contact is between 1500 and 5000 ohms, depending upon the contact resistance. Thus, the applied voltage between points of contact ranges from 0·75 to 2·5 V. It is estimated that a current of 1 mA may be caused to flow in the body of humans by an electric field intensity (at 60 Hz) of 1 MV cm^{-1} (one million volts per centimetre).[120]

In the microwave region, sensation of heat in man occurs between 1 mW cm^{-2} (61 V m^{-1}) and 10 mW cm^{-2} (190 V m^{-1}).[121]

6

Overall Critique and Discussion of the Deleterious Effects of Electric and Magnetic Fields

'Where there is smoke, there is usually fire', so the saying goes.

It has been documented in Chapter 3 that, without a question of a doubt, therapeutic effects of RF, UHF, and microwave electromagnetic radiation have been observed. These are generally due to heating of the affected areas of the body through the application of diathermy machines. It is probable that even with beneficial results, deleterious side effects, called *contra-indications* by the medical profession, may exist. Also in Chapter 3, various examples of 'athermal' therapeutic effects are cited. These are of more questionable authenticity than those obtained by diathermy, primarily because of inconsistency in the duplication of results, even by the same experimenters and therapists. However, because of the multitude and diversity of these claims, they cannot be completely discounted. In fact, the probability of 'athermal' therapeutic effects is enhanced by a number of physiological phenomena which are induced by electromagnetic fields. Some are described in Chapter 4, and are summarised below.

Phosphenes[94]: 800 G, 30–40 Hz.
Radiosounds[96–98]: 250 mW cm^{-2}, peak; 1–7 mW cm^{-2}, average; 200–3000 MHz, modulated.
Cardiovascular effects[99]: 1 mW cm^{-2} (UHF); 10–100 V m^{-1}, short wave, and 100–1000 V m^{-1} (medium wave).
Conditioned vascular reflex[101]: 10 000 V m^{-1} (high frequency).
Reaction time effects[104]: 5–17 G (d.c.) modulated at 0·1–0·2 Hz.

A stumbling block in the acceptance of 'athermal' effects has been the lack of complete understanding of the mechanisms of interaction between electric and magnetic fields and biological systems. Schwan[122] has studied the mechanisms of non-thermal effects in some detail. His influence in this field has been appreciable. The conclusions which he has made are as follows:

1. Force-field effects, such as pearl chain formation and orientation phenomena require strong interactions which can occur only at high

field strengths. For micron-sized particles, field-strength levels in excess of 10 000 V m^{-1} are required. Macromolecules require even higher field levels.

2. Direct effects on biological membranes and neurons, as well as macromolecular resonance denaturation, are both unlikely and contrary to present concepts about the electrical properties of membranes and macromolecules.

3. Macromolecular denaturation at high electrical fields is a remote possibility, but fields of the order of 100 000 V m^{-1} are required to yield complete molecular orientation.

In effect, Schwan appears to discount 'athermal' effects.

Frey[97, 123], in a discussion of possible mechanisms of interaction, states that we really do not have a good understanding of how the nervous system functions. He says that our understanding of how information is coded, transferred, and stored in the nervous system is almost nil, and conclusions (such as those made by Schwan), which are based on particular nerve models are really hypotheses, not fact.

Becker[89, 90, 91] has proposed that each cell contains its own data transmission system that does not depend on commands from a central point, so that a cell can react independently to electromagnetic stimuli. He believes that low-level electrical currents and potentials, produced either by direct injection or by rectification and induction from a field, have the capability of bringing about very major biological effects of a very basic nature.

Cesar Romero-Sierra[90, 124] and his co-workers have found that the relationship between the neuron and glia is electromagnetic in nature. The glia is responsible for the formation of myelin when it interacts with the axon. Therefore, external electromagnetic fields could interact with biological systems through this process.

Cleary,[125] in a very detailed consideration of possible mechanisms of low-intensity interaction of microwave and RF radiation, makes the following observations:

1. Present indications are that non-uniformities in the pattern of energy absorption and the consequent temperature variation within a body, may play a significant role in the induction of low-intensity reversible microwave effects.

2. Microscopic or molecular interaction mechanisms which lead to conformational changes in the molecules or molecular assemblages require too high energy requirements, precluding the possibility of these types of effects at low intensities.

3. Thermally-enhanced molecular interactions may occur at relatively low radiation intensities due to localised temperature rises without significant gross body heating.

Nearly all writers on the subject of 'athermal' effects, who personally do not give credence to 'athermal' effects, end their remarks with a statement something like this:

> While no positive indications of criticality exist from a medical viewpoint, it is important that the medical research being done in the USA be continued to add to our knowledge of the electrical aspects of the human body.[33]

or,

> Most western scientists do not accept the Soviet views . . . All the effects . . . mentioned as reported or potential are speculative and so far have not been accepted as valid. Despite this it is proper for HEW (Health, Education and Welfare Department in the USA) . . . to be conservative and keep investigating these speculative effects.[126]

Such statements may be self-serving in two ways: it does not harm the reputation of the product being marketed; nor, does it close the door to research grants and contracts to prove that the product is safe.

It is true that those who disclaim the possibility of non-thermal effects are in the minority, probably because it is not a popular stand, as Appleton states:[112]

> Therefore, it is not surprising that conclusions that microwave oven, microwave diathermy, and military radar are not sources of harmful electromagnetic pollution have been greeted with a rather tepid response.

In fact, in today's emphasis on the environment and all things ecological, the majority of people, both knowledgeable and unknowledgeable, are very likely to disbelieve this minority. A striking case in point is the Sanguine Project of the US Navy which has encountered extreme difficulty in finding a location where the installation can be made. The northern part of my home State of Wisconsin was selected by the US Navy as a suitable site. Extensive tests were made over a period of three to four years to demonstrate that there would be no interference problems and no adverse environmental effects. For example, an evaluation of the health of personnel working near the Wisconsin test facility showed no evidence of medical problems.[50] Normal 60 Hz electric fields due to power lines in a rural or forest area were measured at 200 points; the maximum electric field was $0{\cdot}040\ \mathrm{V\ m^{-1}}$ and the mean value was $0{\cdot}003\ \mathrm{V\ m^{-1}}$. At 200 points located near consumer locations, the maximum value of electric field was $0{\cdot}500\ \mathrm{V\ m^{-1}}$, and the mean value was $0{\cdot}090\ \mathrm{V\ m^{-1}}$. The estimated level due to the Sanguine Project, operating at full power, is 0·07 V m^{-1} and 0·20 G.[48] However, because of the opposition of the majority of people in northern Wisconsin, the project has been moth-balled in favour of a limited surface installation, called Seafarer, at either the Nellis Air Force Base in Nevada, or at the Fort Bliss (Texas)–White Sands (New Mexico) military complex. I suppose that the majority of people were, and still are, afraid of the unknown. I guess there is a latent fear in all of us about what is new, particularly if there are some scientific adversaries (expert witnesses) to oppose the innovation or new project. A similar result occurred with regard to the supersonic transport programme in the USA. I believe it was scuttled not because of the projected noise it was to thrust upon us in a corridor along its flight path, but because of the unknown extent of its effect on the ozone layer in the upper atmosphere. 'Expert' witnesses predicted dire consequences for all of humankind if the supersonic transports proliferated. After four years of study, these effects have been found to be so small as to be negligible.

In the scientific community, as opposed to the lay public, I believe that there are two forces acting—one objective, the other subjective. On the one hand, objectively we realise that we have been immersed in electromagnetic fields for a considerable length of time with apparently no ill-effects; subjectively, we realise that it is in the realm of possibility that there are effects, and possibility adverse effects. The degree of our scientific background, and specifically our knowledge of the question of electrical pollution, probably governs the relative weight given to each of the viewpoints.

As far as the lay public is concerned, the major force acting is the subjective one. This force is usually strengthened, rather than weakened, by reports published in the popular press and by publications of consumer protection organisations and agencies. Subjectively, the lay public hopes that there have been no adverse effects, but a fear still remains that there may have been adverse effects which will shorten their lives and the lives of their families.

But, are there non-thermal effects due to electric and magnetic fields? If the answer is yes, do these non-thermal effects occur over the entire frequency spectrum, or do they occur only in certain frequency bands? I believe that *there are non-thermal effects*, but not over the whole gamut of frequencies, nor for all the types of fields, covered by the sources described in Chapter 2. What I will state subsequently is, in all probability, not all-inclusive. There may be other mechanisms of interaction, but I believe that what I am setting out is the primary means by which electromagnetic fields interact with biological systems.

Man is not a passive entity. He is active; his heart beats, his blood and other body fluids flow, his nerves conduct impulses to and from the brain, his muscles act to give effect to his every movement. In all this, electric voltages and currents are generated by the myriads of cells which make up every part of man. This is called bioelectricity or electrophysiological phenomena. Three manifestations of bio-generated electricity are: signals from the brain which are measured by the electroencephalograph (EEG); signals from muscles, which are measured by the electromyograph (EMG); signals from the heart, which are measured by the electrocardiograph (ECG). These signals may also be measured in terms of current, for example by the magnetocardiograph (MCG), which can be readily measured today with a superconducting-type of magnetometer.[127]

6.1 EEG

Electroencephalograms are widely used by neurologists, in particular, to diagnose various physiological disorders of the brain as well as various psychological states of the mind. EEG graphs are ordinarily read directly by the neurologists, but, increasing use is being made of automatic computer analysis. The frequency spectrum of the EEG ranges from zero to about 50 Hz. The spectrum is subdivided into discrete bands, with each band predominant in various physiological and psychological states. These are identified in Figure 6.1.[144]

6.2 EMG

Electromyographic potentials are generated by muscle fibres. Needle-type or surface electrodes are used to record EMG signals. Ordinarily, signals from a number of muscle fibres are detected. The frequency components of EMG are between 100 and 3000 Hz. Applications of electromyograms are in dentistry, and measurement of speech, eye movements and breathing.[145, 146] EMG biofeedback has been utilised to relieve tension headaches.[144]

6.3 ECG (MCG)

The electrocardiograph is probably the best known and the largest bioelectric signal of the ones we are describing. ECG can probably be considered as a special case of EMG, since the ECG signals are obtained from muscle fibres and nerves of the heart. Figure 6.2 shows

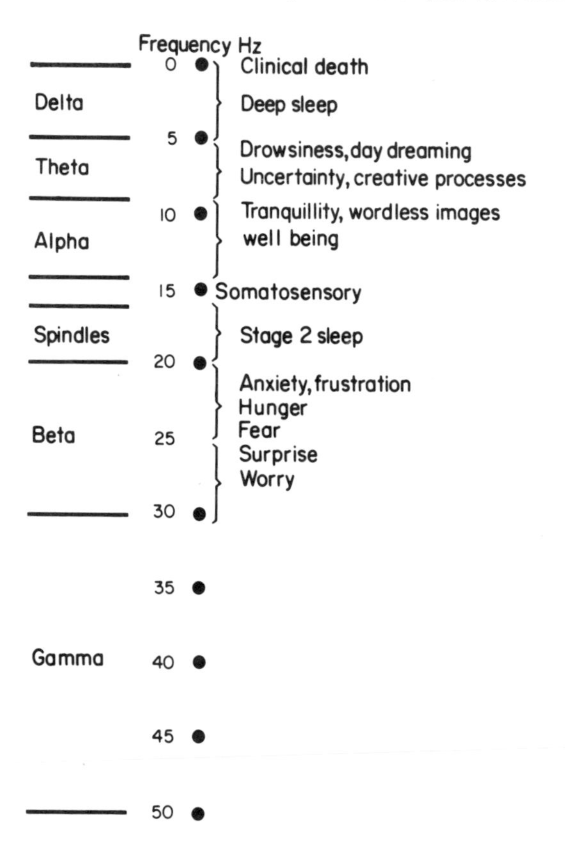

FIGURE 6.1 Electroencephalogram (EEG) spectrum. Two stages of sleep are shown, as well as various physiological and psychological states

the essential parts of the cardiac signal. Each portion is identified by the letters P, Q, R, S, and T. Their sources are as follows:

P wave: contraction of the atria (depolarisation).
PR interval: time taken for conduction of P wave excitation to spread from the sinoatrial (SA) node to the atrioventricular (AV) node.
QRS complex: contraction of the ventricles (depolarisation).
T wave: return of the ventricular mass to the resting electrical state (repolarisation).

Normal heart rate is from 60–100 beats per minutes ($1–1\frac{2}{3}$ Hz). However, the complex nature of the cardiac signal causes the frequency spectrum to extend to much higher values. The greatest portion of the ECG signal occupies a frequency range of between 0·2 and 100 Hz, but frequencies above 100 Hz may occur in some abnormal states.

Normally all muscle fibres of the atria or of the ventricles contract synchronously. When individual muscle fibres contract independently, they are said to be 'fibrillating'. Fibrillation can occur in either the atrial or ventricular muscle. Fibrillation can be caused by excessive externally applied current, which was mentioned previously (Sections 4.1.6 and 5.1.3).

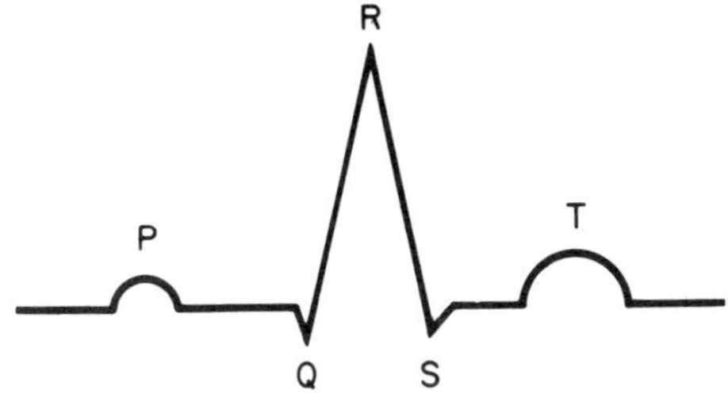

FIGURE 6.2 The cardiac signal, showing the five distinct parts, identified by P, Q, R, S, and T

Magnetocardiographs are obtained by sensing the magnetic fields generated by currents which flow during the depolarisation and repolarisation processes.

A very important research effort has been undertaken at the Brain Research Institute, University of California, Los Angeles (U.C.L.A.) on the effects of biological frequencies (1–25 Hz) on the behaviour and EEG of animals. Reliable changes in behaviour (shifts towards shorter inter-response time) and in EEG (increased rhythmic activity at 6–8 Hz) were observed in monkeys exposed to electric fields (10 V m^{-1}, 7 Hz) over long periods of time. These effects were not observed at 10 Hz.[147]

Untrained cats with chronically implanted electrodes in the brain were tested with very high frequency (147 MHz), low intensity (20 V m^{-1} or less, 1 mW cm^{-2}) amplitude-modulated signals. Amplitude modulation over a wide, low frequency range (0–20 Hz) was used. The results indicated enhanced somnolence under exposures at 1–5 Hz modulation, fast EEG and higher behavioural activities under 16–20 Hz modulation, and no effects at all with unmodulated VHF fields.[148] EEG rhythms which were not in the frequency range of the modulations were not affected by the field exposures. Bawin[148] conjectures that the EEG changes induced by the modulated fields could reflect true neuronal phenomena in the central nervous system.

Frey[90, 123] agrees with the modulation technique. He believes that pulsed current is more effective, and that one must think in terms of modulation, voltage, and frequency.

Phosphenes[94] are caused by alternating magnetic fields of between 30 and 40 Hz. Radiosounds[96–98] are caused by radio frequency modulated by a narrow pulse at a typical rate of 50 Hz.

Electrical current is most dangerous in the frequency range 10–200 Hz.[119]

Various forms of electrotherapy, such as, electroacupunture, electro-sleep, electro-anaesthesia, the transcutaneous relief of chronic pain by stimulation, and healing of wounds and bone and tendon fractures, all employ, either directly or through modulation of a radio frequency carrier, a low frequency of the order of 100 Hz.[77, 78, 91, 92, 93, 124]

Geophysically, it is more than of passing interest to realise that Schumann resonances which are observed on earth occur in the range 5–40 Hz, with several spectral peaks occurring at 8, 13·9, 19·6, 25·4, and 31 Hz.[11] The Schumann resonances may be explained in terms of standing waves which exist in the earth-ionosphere cavity; they are usually excited by thunderstorms. A rather interesting hypothesis has been proposed regarding resonances of this type. Cole and Graf[149] postulate that the emergence and evolution of prebiotic proteins (similar to Sidney Fox's cell-like 'protenoids')[150] were enhanced by an ultra-sensitive protein transreceiver bio-communication mechanism tuned to Schumann-type resonances in primordial earth. They envisage an ELF (extra low frequency) oscillator with a 10 Hz eigenfrequency, whose tuned circuit was the molten core-ionosphere cavity. They conclude,

> Such phenomena as the 10 Hz tonus vibration of muscle, the feeling of wellbeing and enhanced reception during periods of increased alpha-wave activity, an extrinsic timing device for biological rhythms, etc. can be potentially viewed as manifestations of the same mechanism.

Becker concurs with the idea of a cellular communication system[90, 91] and links it to a solid-state mechanism. He believes that each cell contains its own data transmission system that does not depend on commands from a central point. It is commonly known that solid-state devices can serve as demodulators for modulated signals. One could also logically reason in reverse: research at the Brain Research Institute (U.C.L.A.) has demonstrated that modulated signals, in contrast to unmodulated signals, affected the EEG of cats. Only a device having non-linear properties can demodulate a modulated wave; this device could possess the properties of a semi-conductor.

6.4 Conclusion

Yes, thermal electrical pollution does undoubtedly exist. Everyone is in agreement with this statement.

Yes, non-thermal electrical pollution exists. More and more knowledgeable people are beginning to hold this view. Non-thermal electrical pollution is most severe when waves modulated by frequencies which coincide with biological frequencies are used. Fortunately, there are many more cases of electromagnetic radiation which do not fall into this category. Much more research is required to verify this conclusion, and to delineate those types of non-thermal pollution which are to be avoided.

In our technological age, we cannot avoid being immersed in a sea of electrical, magnetic and electro-magnetic fields. But, we can avoid being affected *thermally* by high intensity fields, and *athermally* by the more harmful types of low-intensity fields. We must avoid these hazardous fields just as we must avoid harmful water, air, noise, and thermal pollution. Otherwise, we may find that life expectancy will cease to increase, and, in fact, may begin to decrease due to the unseen pollution—electrical pollution.

References

1. R. W. Rand, S. J. St. Lorant and E. Tillman (1973). 'A superconducting magnet system for surgical applications', Paper 22–8, *Digest of the Intermag Conference*, Washington, D.C., USA
2. D. E. Beischer (1964). 'Survival of animals in magnetic fields of 140,000 oersted', in *Biological Effects of Magnetic Fields*, Vol. 1, M. F. Barnothy (Editor), Plenum Press, New York, pp. 201–208
3. D. E. Beischer (1969). 'Vectorcardiogram and aortic blood flow of squirrel monkeys (*Saimiri sciureus*) in a strong superconductive electromagnet', in *Biological Effects of Magnetic Fields*, Vol. 2, M. F. Barnothy (Editor), Plenum Press, New York, pp. 241–259
4. R. M. Broadhead (1973). 'Thefts', *New Library World*, **74,** 236
5. O. C. Hood, J. M. Keshishian, N. P. D. Smith, E. Podolak, A. A. Hoffman, and N. R. Baker (1972). 'Anti-hijacking efforts and cardiac pacemakers: report of a clinical study', *Aerospace Medicine*, **43,** 314–322
6. D. R. Kelland (1973). 'High gradient magnetic separation applied to mineral benefication', *IEEE Trans. on Magnetics, MAG-9*: 307–310, September
7. S. C. Trinidade and H. H. Kohn (1973). 'Magnetic desulfurization of coal', *IEEE Trans. on Magnetics, MAG-9*: 310–313, September
8. C. de Latour (1973). 'Magnetic separation in water pollution control', *IEEE Trans. on Magnetics, MAG-9*: 314–316, September
9. E. N. Parker (1971). 'Universal magnetic fields', *Am. Sci.,* **59,** 578–585
10. R. Juergens (1972). 'Reconciling celestial mechanics and Velikovskian catastrophism', *Pensee*, **2,** 6–12
11. C. Polk (1974). 'Sources, propagation, amplitude and temporal variation of extremely low frequency (0–100 Hz) electromagnetic fields', in *Biologic and Clinical Effects of Low-Frequency Magnetic and Electric Fields*, J. G. Llaurado, A. Sances, Jr. and J. H. Battocletti (Editors), Chapter 2, C. C. Thomas, Springfield, Illinois
12. A. S. Presman (1970). *Electromagnetic Fields and Life*,

translated by F. L. Sinclair, Edited by F. A. Brown, Jr., Plenum Press, New York, Chapter 11

13. C. C. Conley (1969). 'Effects of near-zero magnetic fields upon biological systems', in *Biological Effects of Magnetic Fields*, Vol. 2, M. F. Barnothy (Editor), Plenum Press, New York
14. D. E. Beischer (1965). 'Biomagnetics', *Ann. N.Y. Acad. Sci.*, **134,** 454
15. D. E. Beischer, E. F. Miller, II, and J. C. Knepton, Jr. (1967). 'Exposure of man to low intensity magnetic fields in a coil system', NAMI-1018, NASA R-39, Naval Aerospace Medical Institute, Pensacola, Florida
16. A. S. Presman (1970). *Electromagentic Fields and Life*, translated by F. L. Sinclair, Edited by F. A. Brown, Jr., Plenum Press, New York, Chapter 2
17. 'Final report completed on solar magnetic disturbances research project', *EEI* (Edison Electric Institute) *Bulletin,* **41,** 233–238, 246, Sept.–Oct. 1973
18. R. Reiter (1964). 'Atmospheric electricity and natural radioactivity', in *Medical Climatology*, Edited by Sidney Licht, Elizabeth Licht (Publisher), New Haven, Connecticut, Chapter 10
19. I. Pavlik 'Significance of air ionization', *Ibid.*, Chapter 11
20. J. B. Beal (1974). 'Electrostatic fields, electromagnetic fields, and ions—mind/body/environment inter-relationships', in *Biologic and Clinical Effects of Low-Frequency Magnetic and Electric Fields*, J. G. Llaurado, A. Sances, Jr. and J. H. Battocletti (Editors), Chapter I, C. C. Thomas, Springfield, Illinois
21. S. Sugiyama. 'Control of visual fatigue by means of d.c. and a.c. electric fields', *Ibid.*, Chapter 5
22. G. D. Friedlander (1972). 'Is "power to the people" going underground?', *IEEE Spectrum*, **9,** 62–71
23. U. Lamm (1966). 'Long-distance power transmission', *International Science and Technology*, 42–48, April
24. F. E. Dominy (1969). 'Economic aspects of the Pacific northwest-southwest intertie', *IEEE Spectrum* **6,** 65–71
25. V. I. Popkov (1969). 'EHV transmission in the Soviet Union', *IEEE Spectrum*, **6,** 18–21
26. M. Klein, J. Linnenkohl and H. Heinderreich (1974). 'HVDC to illuminate darkest Africa', *IEEE Spectrum*, **11,** 51–58
27. J. C. Keesey and F. S. Letcher (1970). 'Human thresholds of electric shock at power transmission frequencies', *Arch. Environ. Health*, **21,** 547–552
28. G. J. Berg (1972) 'Measurement of electrostatic potential below HV lines', *IEEE Trans. Instrum. Meas.*, **IM–21**, 287–288
29, M. L. Singewald, O. R. Langworthy and W. B. Kouwenhoven

(1973) 'Medical follow-up study of high voltage linemen working in a.c. electric fields', *IEEE Trans. Power Appar. Syst.* **PAS–92,** 1307–1309

30. W. B. Kouwenhoven, C. J. Miller, Jr., H. C. Barnes, J. W. Simpson, H. L. Rorden and T. J. Burgess (1966). 'Body currents in live line working', *IEEE Trans. Power Appar. Syst.*, **PAS–85(4),** 403–411
31. T. P. Asanova and A. N. Rakov (1966). 'The health status of people working in the electric field of open 400–500 kV switching structures', *Gigiena Truda i Professional nye Zabolevaniia*, **10,** 50–52
32. *Standards and Regulations for the Protection of Labor during Work on Substations and Transmission Lines at Voltages of 400-, 500-, and 750-KV Alternating Current at Industrial Frequency*, Ministry of Power Engineering and Electrification of the USSR and Ministry for the Protection of Health of the USSR, Moscow 1972. Translated by the Engineering Reference Branch, Denver, Colorado, July 1973; Translation No. 393
33. H. C. Barnes and B. Thoren (1974), 'UHV: It's only a matter of time', *Electr. Light Power*, T/D 9–13
34. E. D. Sunde (1968) *Earth Conduction Effects in Transmission Systems*, Dover
35. A. R. Valentino (1974). 'Laboratory simulation of extremely low frequency electric and magnetic fields', in *Biologic and Clinical Effects of Low-Frequency Magnetic and Electric Fields*, J. G. Llaurado, A. Sances, Jr. and J. H. Battocletti (Editors), Chapter 12, C. C. Thomas, Springfield, Illinois
36. D. A. Miller (1974). 'Electric and magnetic fields produced by commercial power systems', *Ibid.*, Chapter 4
37. V. P. Korobkova, Yu. A. Morozov, M. D. Stolarov and Yu. A. Yakub (1972). 'Influence of the electric field in 500 and 750 kV switchyards on maintenance staff and means for its protection', *Int. Conf. on Large High Tension Electric Systems* (CIGRE), Paper 23–06, Paris
38. G. D. Friedlander. Various articles in *IEEE Spectrum*, **4,** 62–67, 1967; **5,** 50–65, 1968; **5,** 56–66, 1968; **5,** 77–90, 1968; **9,** 63–66, 1972; **9,** 34–46, 1972; **9,** 60–72, 1972; **9,** 41–54, 1972; **11,** 53–64, 1974; and **11,** 62–70, 1974
39. Special Issue on Transportation, *Proc. IEEE*, **56,** 377–729, April 1968
40. Special Issue on Ground Transportation for the Eighties, *Proc. IEEE*, **61,** 516–656, May 1973
41. J. A. Ross (1973). 'ROMAG transportation system', *Proc. IEEE*, **61,** 617–620

42. R. D. Thornton (1973). 'Flying low with Maglev', *IEEE Spectrum*, **10,** 47–54
43. E. Ohno, M. Iwamoto and T. Yamada (1973). 'Characteristics of superconductive magnetic suspension and propulsion for high-speed trains', *Proc. IEEE*, **61,** 579–586
44. Y. Iwasa (1973). 'Magnetic shielding for magnetically levitated vehicles', *Proc. IEEE*, **61,** 598–603
45. J. T. Sahili (1975). 'Kilowatthours *vs.* liters', *IEEE Spectrum*, **12,** 62–66
46. T. J. Healey (1974). 'The electric car: will it really go?', *IEEE Spectrum*, **11,** 50–53, and **11,** 32, 36, and **11,** 24, for discussions
47. Special Issue on Project Sanguine, *IEEE Trans. Commun.,* April 1974
48. R. E. Baker (1974). 'Project Sanguine: Overview and status of the Navy's ELF communications system concept', in *Biologic and Clinical Effects of Low-Frequency Magnetic and Electric Fields*, J. G. Llaurado, A. Sances, Jr. and J. H. Battocletti (Editors), Chapter 6, C. C. Thomas, Springfield, Illinois
49. T. C. Rozzell (1974). 'Biological research for extremely low frequency communication systems', *Ibid.*, Chapter 7
50. P. E. Krumpe and M. S. Tockman (1974). 'Evaluation of the health of personnel working near Project Sanguine Beta test facility from 1971 to 1972', *Ibid.*, Chapter 8
51. L. F. Mills and P. Segal (1970). *Radiation Incidents Registry Report 1970*, Report BRH/DBE 70–6, U.S. Department of Health, Education, and Welfare, Public Health Service, Bureau of Radiological Health, Washington, D.C., USA
52. D. I. Weinberg (1970). 'How the electric-shock hazard arises', in *Electric Hazards in Hospitals. Proceedings of a Workshop*, pp. 11–23, National Academy of Sciences, Washington, D.C., USA
53. R. A. Tell (1972). 'Broadcast radiation: how safe is safe?', *IEEE Spectrum*, **9,** 43–51
54. S. W. Smith and D. G. Brown (1971). *Radio Frequency and Microwave Radiation Levels Resulting from Man-made Sources in the Washington, D.C. Area*, Publication No. FDA 72–8015 and BRH/DEP 72–5, U.S. Dept. of Health, Education and Welfare, Public Health Service, Food and Drug Administration, Bureau of Radiological Health, Washington, D.C., USA
55. K. R. Envall, R. W. Peterson and H. F. Stewart (1971). *Measurement of Electromagnetic Radiation Levels from Selected Transmitters Operating Between 54 and 220 MHz in the Las Vegas, Nevada, Area*, Publication No. (FDA) 72–8012, BRH/Dep 72–4, USDHEW, Public Health Service, Food and

Drug Administration, Bureau of Radiological Health, Washington, D.C., USA

56. J. Y. Harris (1970). *Electronic Product Inventory Study*, Publication No. BRH/DEP 70–29, U.S. Department of Health, Education, and Welfare, Public Health Service, Washington, D.C., USA
57. 'Bouncing beams with a bubble-in-the-sky' (1975). *Ind. Res.*, **17(3),** 26, 28
58. 'Finger-tip halos of Kirlian photography' (1973). *Med. World News*, pp. 43–48
59. 'It's a tree . . . . a pole . . . . a man; No! A short-range hf antenna', *Electron. Des.*, **21,** 52, 54, 1973 and **22,** 7, 9 1974
60. S. K. Ghosh (1973). 'Safety considerations for microwave diathermy units' in *Health Physics in the Healing Arts—Health Physics Society Seventh Mid-year Topical Symposium*, DHEW Pub. (FDA) 73–8029, pp. 479–486
61. *A Partial Inventory of Microwave Towers, Broadcasting Transmitters, and Fixed Radar by States and Regions*, Pub. No. BRH/DEP 70-15, 1970, U.S. Department of Health, Education and Welfare, Public Health Service, Washington, D.C., USA
62. R. E. Fenton and K. W. Olson (1969). 'The electronic highway', *IEEE Spectrum*, **6,** 60–66
63. A. S. Clorfeine (1973). 'Driving under the influence of electronics', *IEEE Spectrum*, **10,** 32–37
64. J. Shefer and R. J. Klensch (1973). 'Harmonic radar helps autos avoid collisions', *IEEE Spectrum*, **10,** 38–45
65. 'A microwave car license plate designed, with a spate of uses' (1974). *Electron. Des.,* **22,** 36
66. R. L. Elder, J. A. Eure, and J. W. Nicolls (1974). 'Radiation leakage control of industrial microwave power devices', *J. of Microwave Power*, **9,** 51–61
67. L. C. Seabron and L. W. Coopersmith (1971). *Results of the 1970 Microwave Oven Survey*, Pub. No. FDA 72–8007 and BRH/DEP 72–2, USDEW, Public Health Service, Food and Drug Administration, Bureau of Radiological Health, Washington, D.C., USA
68. N. N. Hankin (1974). *An Evaluation of Selected Satellite Communication Systems as Sources of Environmental Microwave Radiation*, U.S. Environmental Protection Agency, Report No. EPA–520/2–74–008
69. D. A. Dunn (1968). 'Power from microwaves', *Sci. Technol.*, No. 80, 26–37
70. T. K. Ishii (1968). 'Development of a self-starting microwave motor', *J. Microwave Power*, **3(3),** 134–142

71. W. C. Brown (1973). 'Satellite power stations: a new source of energy?', *IEEE Spectrum*, **10,** 38–47
72. W. C. Brown (1973).'Adapting microwave techniques to help solve future energy problems', *IEEE Trans. Microwave Theory Tech.*, **MTT-21,** 753–763
73. Yu. A. Kholodov (1970). *Magnetism in Biology*, 'Science' Publishing House, Academy of Science, Moscow, USSR (in Russian)
74. Yu. A. Kholodov (1972). *Man in the Magnetic Web (The Magnetic Field and Life)*, Znaniye Publishing House, Moscow. (An English translation is being made in the USA at the request of J. B. Beal, Miami Heart Institute, Miami Beach, Florida)
75. A. L. Watkins (1968). *A Manual of Electrotherapy*, 3rd Ed., Lea and Febiger, Philadelphia
76. S. Licht (Editor) (1959). *Therapeutic Electricity and Ultraviolet Radiation*, Elizabeth Licht (Publisher), New Haven, Conn.
77. J. G. Llaurado, A. Sances, Jr. and J. H. Battocletti (Editors) (1974). *Biologic and Clinical Effects of Low-Frequency Magnetic and Electric Fields*, C. C. Thomas, Springfield, Illinois, Part Three and Workshop 3
78. *Electrical, Surgical and Pharmacological Management of Pain*, Proceedings of the Neuroelectric Society, Seventh Annual Meeting, New Orleans, Louisiana, November 20–23, 1974
79. G. W. de la Waar and D. Baker (1967). *Biomagnetism*, Delawarr Laboratories Ltd, Oxford, England
80. S. Zurkerman (1973). 'The Great Bordeaux magnetic machine mystery', *Sunday Times Weekly Review (London)*, January 7, 1973, pp. 25–26
81. D. S. Greenberg. 'The French Concoction', *Saturday Review of The Sciences*, May–Aug. 1973, pp. 36–44
82. M-R Rivière, A. Priore, F. Berlureau, M. Fournier, and M. Guérin. 'Action de champs électromagnétiques sur les greffes de la tumeur T 8 chez le rat', *C.R. Acad. Sc. Paris*, **259,** 4895–4897, Group 14, 21 December 1964
83. M-R Rivière and M. Guèrin. 'Nouvelles recherches effectuées chez des rats porteurs d'un lymphosarcome lymphoblastique soumis à l'action d'ondes électromagnétiques associées à des champs magnétiques', *C.R. Acad. Sc. Paris*, **262,** 2669–2672, Série D, 20 June 1966
84. M-R Rivière, A. Priore, F. Berlureau, M. Fournier and M. Guérin. 'Phenomènes des régression observés sur les greffes d'un lymphosarcome chez des souris exposées á des champs électromagnétiques', *C.R. Acad. Sc. Paris*, **260,** 2639–2643, Groupe 14, 1 March 1965

85. R. Pautrizel, A. Priore, F. Berlureau, and A-N Pautrizel. 'Action de champs magnétiques combinés á des ondes électro-magnétiques sur la trypanosomose expérimentale du lapin', *C.R. Acad. Sc. Paris*, **271,** 877–880, Série D, 7 September 1970
86. D. G. Remark (1971). *Survey of Diathermy Equipment Use in Pinellas County*, Florida, Report BRH/NERHL 71–1, Superintendent of Documents, US GPO, Washington, D.C., USA
87. J. Daels (1973). 'Microwave heating of the uterine wall during parturition', *Obstet. Gynecol.*, **42(1),** 76–79
88. A. S. Presman (1970). *Electromagnetic Fields and Life*, translated by F. L. Sinclair, Edited by F. A. Brown, Jr., Plenum Press, New York, Chapter 14
89. R. O. Becker (1972). 'Electromagnetic forces and life processes', *Technol. Rev.*, pp 32–28, December issue.
90. 'Clinical applications of electric current remain largely unexplored' (1974). *J. Am. Med. Assoc.*, **227,** 129–130
91. R. O. Becker (1974). 'The basic biological data transmission and control system influenced by electrical forces', *Ann. N.Y. Acad. Sc.*, **238,** 236–241
92. C. A. L. Bassett, R. J. Pawluk, and A. A. Pilla (1974). 'Acceleration of fracture repair by electromagnetic fields. A surgically noninvasive method', *Ann. N.Y. Acad. Sci.*, **238,** 242–262
93. 'Electrical current as a bone healer' (1975). *Med. World News*, **16(2),** 84
94. H. S. Alexander (1962). 'Biomagnetics—the biological effects of magnetic fields', *Am. J. Med. Electron.*, **1,** 181–187
95. A. Kolin. 'Magnetic fields in biology', *Physics Today*, pp. 39–50, November 1968
96. A. H. Frey (1963). 'Some effects on human subjects of ultra-high frequency radiation', *The Am. J. Med. Electron.*, **2,** 28–31
97. A. H. Frey (1971). 'Biological function as influenced by low-power modulated R-F energy', *IEEE Trans. Microwave Theory Tech.*, **MTT–19,** 153–164
98. A. H. Frey and R. Messenger, Jr. (1973). 'Human perception of illumination with pulsed ultra-high frequency electromagnetic energy', *Science*, **181,** 356–358
99. A. S. Presman (1970). *Electromagnetic Fields and Life*, translated by F. L. Sinclair, Edited by F. A. Brown, Jr., Plenum Press, New York, Chapter 7
100. S. M. Michaelson (1974). 'Review of program to assess the effects of man from exposure to microwaves', *J. Microwave Power*, **9,** 147–161
101. A. S. Presman (1970). *Electromagnetic Fields and Life*,

translated by F. L. Sinclair, Edited by F. A. Brown, Jr., Plenum Press, New York, Chapter 5

102. R. O. Becker (1969). 'The effect of magnetic fields upon the central nervous system', in *Biological Effects of Magnetic Fields*, Vol. 2, Edited by M. F. Barnothy, Plenum Press, New York
103. A. D. Pokorny and R. B. Mefferd (1966). 'Geomagnetic fluctuations and disturbed behavior', *J. Nerv. Ment. Dis.*, **143,** 140–151
104. H. Friedman, R. O. Becker, and C. H. Bachman (1967). 'Effect of magnetic fields on reaction time performance', *Nature*, **213,** 949–956
105. R. J. Sclabassi, N. S. Namerow and N. F. Enss (1974). 'Somatosensory response to stimulus trains in patients with multiple sclerosis', *Electroencephalogr. Clin. Neurophysiol.*, **37,** 23–33
106. N. S. Namerow, R. J. Sclabassi and N. F. Enss (1974). 'Somatosensory responses to stimulus trains: normative data', *Electroencephalogr. Clin. Neurophysiol.*, **37,** 11–21
107. J. T. Bigger, Jr. (1970). 'Clinical exposure of patients to electric hazards', in *Electric Hazards in Hospitals. Proceedings of a Workshop*, Editor C. W. Walter, National Acad. of Sciences, Washington, D.C., USA, pp. 29–47
108. J. M. R. Bruner (1970). 'The horror of common practice', *Ibid.*, pp. 119–129
109. M. L. Levison. 'Viewpoint. The only thing we have to fear . . .', *Electron. Eng. Times*, September 10, 1973. An opposing letter by R. J. Krusberg in *Electron. Eng. Times*, October 8, 1973.
110. W. C. Milroy and S. M. Michaelson (1972). 'Microwave cataractogenesis: a critical review of the literature', *Aerosp. Med.*, **43,** 67–75
111. F. G. Hirsch. 'Microwave cataracts—a case report reevaluated', in *Electronic Product Radiation and the Health Physicist*, Pub. No. BRH/DEP 70–26, U.S. Dept. of Health, Education and Welfare, Available at National Technical Information Service, Springfield, Va. 22151.
112. B. Appleton (1974). 'Microwave cataracts', *J. Am. Med. Assoc.*, **229,** 407–408
113. M. M. Zaret (1972). 'Clinical aspects of nonionizing radiation', *IEEE Trans. Biomed. Eng.*, **BME–19,** 313–316
114. W. C. Milroy and S. M. Michaelson (1972). 'Thyroid pathophysiology of microwave radiation', *Aerosp. Med.*, **43,** 1126–1131
115. J. R. Swanson, V. E. Rose and C. H. Powell. 'A review of international microwave exposure guides', in *Electronic Product*

Radiation and the Health Physicist, Pub. No. BRH/DEP 70–26, U.S. Dept. of Health, Education and Welfare, Washington, D.C., USA., pp. 95–110

116. S. M. Michaelson and W. M. Houk. 'Exposure criteria for nonionizing radiant energy in the healing arts', in *Health Physics in the Healing Arts—Health Physics Society Seventh Midyear Topical Symposium*, DHEW Pub. (FDA) 73–8029, March 1973, pp. 463–466
117. K. Marha (1971). 'Microwave radiation safety standards in eastern Europe', *IEEE Trans. Microwave Theory Tech.*, **MTT–19(2),** 165–168
118. Second Report on 'Program for Control of Electromagnetic pollution of the environment: the Assessment of Biological Hazards of Nonionizing Electromagnetic Radiation', Office of Telecommunication Policy, Executive Office of the President (USA), May 1974
119. C. F. Dalziel (1972). 'Electric shock hazard', *IEEE Spectrum*, **9(2):** 41–50
120. G. E. Kaufman and S. M. Michaelson (1974). 'Critical review of the biological effects of electric and magnetic fields', in *Biologic and Clinical Effects of Low-Frequency Magnetic and Electric Fields*, J. G. Llaurado, A. Sances, Jr. and J. H. Battocletti (Editors), C. C. Thomas, Springfield, Illinois, Chapter 3
121. A. S. Presman (1970). *Electromagnetic Fields and Life*, translated by F. L. Sinclair, Edited by F. A. Brown, Jr., Plenum Press, New York, Chapter 4
122. H. P. Schwan (1972). 'Microwave radiation: biophysical considerations and standards criteria', *IEEE Trans. Bio-Med. Electron.* **BME–19(4),** 304–312
123. A. H. Frey (1974). 'Differential biologic effects of pulsed and continuous electromagnetic fields and mechanisms of effect', *Ann. N.Y. Acad. Sci.*, **238,** 273–282
124. C. Romero-Sierra and J. A. Turner (1974). 'Biological effects of nonionized radiation: an outline of fundamental laws', *Ann. N.Y. Acad. of Sci.*, **238,** 263–272
125. S. F. Cleary (1973). 'Uncertainties in the evaluation of the biological effects of microwave and radio-frequency radiation', *Health Phys.*, **25,** 387–404
126. J. M. Osepchuk. 'Dctailed critical review of *Consumer Reports* article on microwave ovens: "Not Recommended"', April, 1973. Unpublished Report, but distributed to various people.
127. W. L. Goodman, V. W. Hesterman, L. H. Rorden and W. S. Goree (1973). 'Superconducting instrument systems', *Proc. IEEE*, **61(1),** 20–27

128. E. H. Frei (1972). 'Biomagnetics', *IEEE Trans. on Magnetics*, **MAG–8,** 407–413
129. N. Carlisle and J. Carlisle (1966). *Marvels of Medical Engineering*, The Oak Tree Press, London and Melbourne
130. R. E. Peck (1974). *The Miracle of Shock Treatment*, Exposition Press, Jericho, New York
131. L. A. Daily (1943). 'A clinical study of the results of exposure of laboratory personnel to radar and high frequency radio', *U.S. Nav. Med. Bull.*, **41,** 1052–1056
132. F. G. Hirsch and J. T. Parker (1952). 'Bilateral lenticular opacities occurring in a technician operating a microwave generator', *AMA Arch. Ind. Hyg. Occup. Med.*, **6,** 512–517
133. C. I. Barron and A. A. Baraff (1958). 'Medical considerations of exposure to microwaves (radar)', *JAMA*, **168,** 1194–1199
134. I. S. Shimkovich and V. G. Shilyaev (1959). 'Cataract of both eyes which developed as a result of repeated short exposures to an electromagnetic field of high density', *Vestn. Oftal.*, **72,** 12–16
135. L. Mineckie (1961). 'The health of persons exposed to the effect of high frequency electromagnetic fields', *Medycyna Pracy* (Poland), **12,** 337–344 (FTD–TT–61–380)
136. M. Zaret, S. Cleary, B. Pasternack and M. Eisenbud (1961). *Occurrence of Lenticular Imperfections in the Eyes of Microwave Workers and their Association with Environmental Factors*, New York University Press, New York (RADC–TN–61–226)
137. S. E. Belova (1962). 'The effects of microwave irradiation on the eye' in *The Effects of Radar on the Human Body* (Results of Russian studies on the subject), Editor J. J. Turner, pp. 43–48, AD 278172
138. S. F. Cleary, B. S. Pasternack and G. W. Beebe (1965). 'Cataract incidence in radar workers', *Arch. Environ. Health*, **11,** 179–182
139. S. F. Cleary and B. S. Pasternack (1966). 'Lenticular changes in microwave workers. A statistical study', *Arch. Environ. Health*, **12,** 23–29
140. G. H. Kurz and R. B. Einaugler (1968). 'Cataract secondary to microwave radiation', *Am. J. Ophthamol.*, 66, 866–867
141. K. Majewska (1968). 'Investigation on the effect of microwaves on the eye', *Polish Med. J.*, **VII,** 989–994
142. M. Zaraet (1968). 'Ophthalmic hazards of microwave and laser environments', *39th Ann. Sci. Meeting Aerospace Med. Assoc.*, Miami, Florida
143. M. Zaret (1969). 'Ophthalmic hazards of microwave and laser environments', *40th Ann. Sci. Meeting Aerospace Med. Assoc.*, San Francisco, Calif.

144. L. Birk (1973). *Biofeedback Behavioral Medicine*, Grune and Stratton, New York, p. 23
145. E. E. Suckling (1961). *Biolectricity*, McGraw-Hill, New York
146. J. L. Trimble, B. L. Zuber and S. N. Trimble (1973). 'A spectral analysis of single motor unit potentials from human extraocular muscle', *IEEE Trans. on Biomedical Engineering, BME*-**20**(2), 148–151
147. R. J. Gavalas, D. O. Walter, J. Hamer and W. R. Adey (1970). 'Effect of low level, low frequency electric fields on EEG and behaviour in *Macaca nemestrina*', *Brain Research*, **18,** 491–501
148. S. M. Bawin, R. J. Gavalas-Medici and W. R. Adey (1974). 'Reinforcement of transient brain rhythms by amplitude-modulated VHF fields', in *Biologic and Clinical Effects of Low-Frequency Magnetic and Electric Fields*, J. G. Llaurado, A. Sances, Jr. and J. H. Battocletti (Editors), C. C. Thomas, Springfield, Illinois, Chapter XIII
149. F. E. Cole and E. R. Graf (1974). 'Extra low frequency (ELF) electromagnetic radiation as a biocommunications medium: A protein transreceiver system', Ibid., Chapter X
150. S. Fox (1968). 'How did life begin?', *Science and Technology*, February

Index